William Osler

Lectures on the Diagnosis of abdominal Tumors

Delivered to the post-graduate Class Johns Hopkins University, 1893

William Osler

Lectures on the Diagnosis of abdominal Tumors
Delivered to the post-graduate Class Johns Hopkins University, 1893

ISBN/EAN: 9783337177607

Printed in Europe, USA, Canada, Australia, Japan

Cover: Foto ©berggeist007 / pixelio.de

More available books at **www.hansebooks.com**

LECTURES ON THE

DIAGNOSIS OF ABDOMINAL TUMORS

DELIVERED TO THE POST-GRADUATE CLASS
JOHNS HOPKINS UNIVERSITY, 1893

BY

WILLIAM OSLER, M. D.
PROFESSOR OF MEDICINE, JOHNS HOPKINS UNIVERSITY
PHYSICIAN-IN-CHIEF, JOHNS HOPKINS HOSPITAL, BALTIMORE

*REPRINTED FROM
THE NEW YORK MEDICAL JOURNAL, 1894*

NEW YORK
D. APPLETON AND COMPANY
1896

"THE sources to which we turn for evidence respecting the existence and nature of abdominal tumors are: the form and appearance presented to the eye; the form still further discovered by the touch; the resistance ascertained by pressure; the sounds elicited by percussion; and, in a few instances, the sounds perceptible to the ear, either alone or by the aid of the stethoscope; and besides these local and physical signs, we look to the general condition of the system, and of the various excretions, as rendering us most important assistance, and being frequently indispensable toward the formation of a tolerably correct diagnosis."—(Richard Bright, *On Abdominal Tumors.*)

CONTENTS.

LECTURE	PAGE
I. Tumors of the Stomach	1
I. Tumor formed by Dilated Stomach	2
II. Tumor formed by Contracted Stomach	31
II. Nodular and Massive Tumors of the Stomach	35
(a) Tumors of the Pyloric Region	36
(b) Tumors of the Body of the Stomach	52
(c) Massive Tumors of the Stomach	57
III. Tumors of the Liver	67
I. Tumor formed by the Liver itself	68
II. Abscess	70
III. Syphilis	81
IV. Cancer	88
IV. Tumors of the Gall Bladder	99
(a) Dilated Gall Bladder	99
(b) Ill-defined Nodular Tumors at Liver Edge	112
(c) Cancer of the Gall Bladder	118
V. Miscellaneous Tumors	124
I. Tumors of the Intestine	124
II. Omental Tumors	135
III. Tumors of the Pancreas	137
IV. Miscellaneous Tumors	142
(a) Cyst of Mesentery	142
(b) Multiple Tumor Masses in Abdomen	148
(c) Uterine Fibroid	150
(d) Sarcoma of the Abdominal Wall	151
(e) Tumors of Doubtful Nature	153
(f) Aneurysm of the Aorta	156
VI. Tumors of the Kidney	160
I. Movable Kidney	160
(a) Errors in Diagnosis of	160
(b) Dietl's Crises in	165
II. Intermittent Hydronephrosis	170
III. Malignant Disease of Kidney	184
IV. Tuberculous Kidney	189

LECTURES ON THE
DIAGNOSIS OF ABDOMINAL TUMORS.

LECTURE I.

TUMORS OF THE STOMACH.

GENTLEMEN: I propose in the following course to bring before you the experience gleaned during a period of twelve months in the cases of abdominal tumor which have come before me for diagnosis. I have not included the cases admitted under the care of Dr. Thayer (my first assistant) during my absence in July and August, unless I had previously or have afterward seen them. The condition has been dictated at the time of examination, the diagnosis made, when possible, and the subsequent history of the cases has been carefully followed. I have not included in the list instances of ascites, appendicitis, or simple enlargement of the liver or spleen; only cases in which a definite tumor existed in connection with one or other of the abdominal organs. We shall take up the cases in the following order: stomach, of which there were twenty-four, liver and appendages, intestines and peritonæum, renal, and miscellaneous.

In the diagnosis of abdominal tumors Bishop Butler's maxim that "probability is the rule of life" is particularly true, and the cocksureness of the clinical physician, who formerly had to dread only the mortifying disclosures of the post-mortem room, is now wisely tempered when the surgeon can so promptly and safely decide upon the nature of an obscure case.

With the methods of examination of the stomach you are all familiar, having frequently seen them applied; and as elaborate details are available both in the text-books on physical diagnosis, and more fully in the recent special works on diseases of the stomach by Ewald,* Boas,† Bouveret,‡ Debove, and Rémond,# I shall proceed at once to the consideration of the subject in hand.

Tumors of the stomach are formed (1) by the organ itself in a condition of abnormal dilatation or contraction; (2) by nodular or massive outgrowths of its walls.

I. THE TUMOR FORMED BY A DILATED STOMACH.— There were thirteen cases of dilated stomach in the series, in ten of which the organ itself formed a prominent tumor *visible on inspection*. These will form the subject of the present lecture. In all of the cases the existence of a nodular pyloric tumor was also determined. In another case, not considered here, the dilatation of the stomach was caused by the pressure on the duodenum of a tumor of the colon. I will first read to you the histories of the cases, sometimes with the comments dictated at the time of examination, and then make some general remarks on the diagnosis of dilated stomach. Though the condition is common, I am surprised that general practitioners so frequently overlook its presence, owing in large measure to the transgression of one of the primary rules of diagnosis, namely, to carefully and systematically go through

---

* *Klinik der Verdauungskrankheiten.* Dritte Auflage. Berlin.
† *Diagnostik und Therapie der Magenkrankheiten.* Theil ii, Leipsic.
‡ *Traité des maladies de l'estomac.* Paris.
# *Traité des maladies de l'estomac.* Paris.

the routine of inspection, palpation, percussion, and inflation.

CASE I. *Tumor caused by Dilated Stomach; Nodular Tumor in Right Epigastrium; Waves of Peristalsis.*—George A., aged thirty-nine, admitted September 1st, complaining of pain in the abdomen and vomiting. Patient is a tailor by occupation, and has used alcohol to excess. Present illness began last Christmas with symptoms of dyspepsia, occasional vomiting, eructations, and pain in the region of the navel. The pain was much worse after eating and was described as of a gnawing character. The food very often turned sour. Has never vomited any blood. Lately the attacks of vomiting have come on at longer intervals and large quantities of brownish, foul-smelling material have been ejected.

Present Condition.—Patient is a medium-sized man, much emaciated, particularly in the trunk and extremities; there are no glandular enlargements. The tongue is thickly furred. The abdomen is flat, somewhat scaphoid, but presents a slight prominence above and to the left of the navel. At intervals of a minute or two there appears in the epigastrium and upper umbilical region a prominent tumor, the longest diameter transverse, and having somewhat the shape of the stomach. The chief prominence is in the left hypochondrium, and the greater curve emerges beneath the costal margin in the left nipple line, passes obliquely downward to about two inches below the level of the navel, and then turns upward and to the right, reaching nearly to the ribs. The lesser curve, not so distinct, passes two inches from the ensiform cartilage. During the prominence of the tumor waves of contraction pass from left to right and there is sometimes a well-marked depression separating the prominent masses to the left and right of the middle line. During the periods of contraction the masses are firm and resistant; in the intervals they almost completely disappear and the abdomen in these regions is quite soft. In the right parasternal line, just below the edge of the liver, is a nodular tumor.

Fig. 1 is from a photograph taken during the passage of the waves of contraction, three of which are plainly to be seen at the situations marked with the crosses. The letter *f* is placed in the depression separating the stomach into right and left. After several attacks of vomiting, and after having the stomach thoroughly washed out, the distention was very much less marked, and the peristaltic movements were less frequent. The nodular tumor mass was then felt to be very much more in the middle line. For

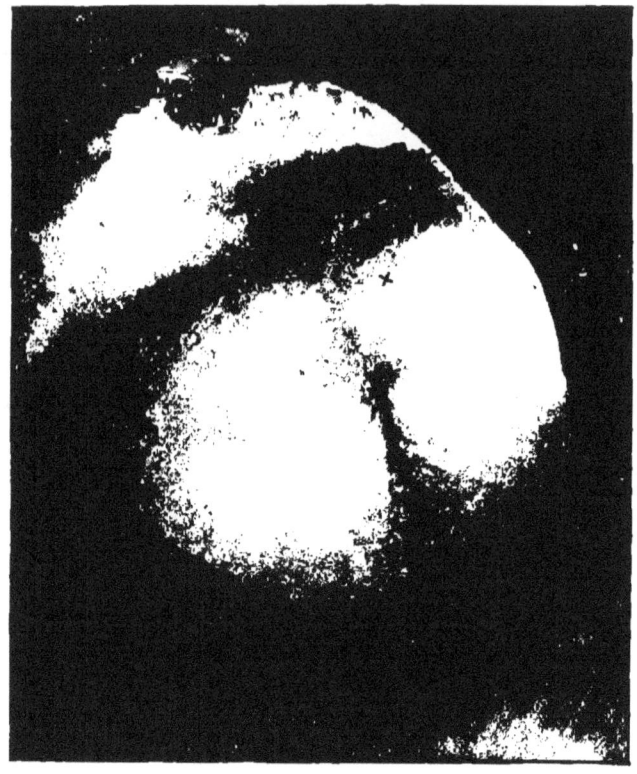

FIG. 1.—From a photograph by Dr. Hewetson, showing undulatory waves of peristalsis in Case I. The crosses are placed on the three prominent waves. The letter *f* indicates the depression on the lesser curve.

a week or ten days before his death this patient had tetany, which is not a very uncommon event in dilatation of the stomach. Death occurred September 26th.

The autopsy showed at the pyloric extremity of the stomach a crater-like tumor mass eight by seven centimetres, the margins thick, elevated, and indurated. Externally there was great thickening about the pylorus, with numerous nodules on the peritonæum. At the pylorus the tumor was massed about the orifice, through which, however, the little finger could pass. The coats of the stomach were enormously thickened. Fig. 2, from a photograph taken on the post-mortem table, shows well the dilatation of the stomach.

CASE II. *Dilated Stomach, forming a Prominent Tumor; Ill-defined Flattened Mass in Right Umbilical Region.*—John L., aged fifty-eight years, seen with Dr. Bryson Wood, September

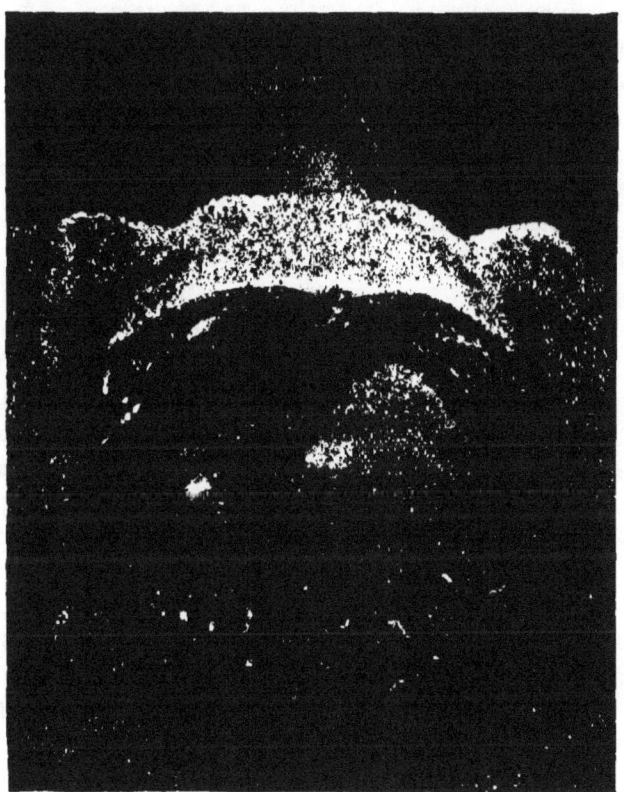

FIG. 2.—Carcinoma of pylorus, showing the dilatation of the stomach as it appeared at autopsy. From a photograph by Dr. Hewetson.

13th, complaining of indigestion and loss of weight. The patient is a tall, large-framed man, who has lived a life of unusual energy and activity, and prior to 1875 had been a hard drinker.

His mother died of some stomach trouble, the precise nature of which he does not know. With this exception, his family history is good. He has always had to be a little careful about eating, but until within the past six months has had good health. The present illness began with dyspeptic symptoms, eructations of gas, feelings of distress a few hours after eating, and occasional vomiting. The chief discomfort was at night, five or six hours after the last meal. Lately these features have increased very much; he has not been able to take solid food; the eructations of gas have become very marked, and he has had at intervals vomiting of large quantities of liquid and undigested food. He has lost rapidly in weight, and has fallen from a hundred and ninety-five to a hundred and forty-two pounds.

The condition on examination was as follows: Large-framed man, not cachectic-looking, moderately emaciated. The tongue has a light white fur.

The abdomen is below the level of the costal margin. In the upper zone, occupying the left epigastric, the left umbilical, and the left hypochondriac regions, there is an irregular swelling which at intervals shows waves of peristalsis and assumes a shape suggestive of a distended stomach. A lesser curvature can be distinctly seen three fingerbreadths from the ensiform cartilage; a greater curvature about two inches below the level of the navel. The most marked prominence is just beneath the left costal margin. To the right the outline of the swelling extends beyond the nipple line. The contrast between the upper and lower abdominal zones is very striking, and the diagnosis of the condition could be made at a glance, as the organ hardened when the waves of peristalsis passed over it.

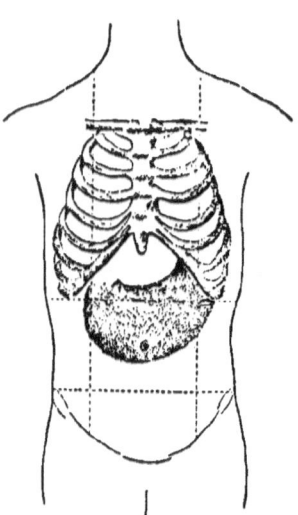

Fig. 3.—Showing the stomach outlines in Case II.

On palpation, the abdomen was everywhere soft and there was no tenderness. During contraction the stomach was firm and resistant. There was no nodular tumor to be felt, although between the navel and the right costal margin there was a sense of increased resistance, particularly beneath the ribs. The area of liver dullness was diminished. There was no enlargement of the superficial glands.

The patient was ordered to have the stomach washed out every morning, and to take a diet of milk and egg-white.

October 17th.—Patient was seen again to-day with Dr. Salzer, partly with a view of determining the advisability of a Loreta's operation. Since the last note the patient has improved considerably under the daily use of the stomach tube, and he has been able to take Leube's beef extract, meat balls, and small quantities of milk without discomfort. He has not, however, gained in weight; still looks very haggard and emaciated, and says he at times feels very queer in his head, as if he would go crazy.

The abdomen is a little full in the upper zone, and every few minutes the distinct outline of the stomach can be plainly seen, forming a tumor of unusual prominence. The stomach tympany can be obtained as high as the fifth interspace in the parasternal line.

On palpation, there is no thickening or nodular mass to be felt in the epigastric region, nor on the deepest inspiration can any mass be felt beneath the left costal margin. Just below the limit of the stomach, and to the right of the navel, there is an ill-defined flattened mass, which does not, however, feel like a thickened pylorus, nor is it likely that the pylorus could be felt in this situation with the stomach tympany and the outline of the stomach passing, as it does to-day, with such distinctness beneath the right costal margin. It seems more probable that the pylorus is covered by the distended organ.

Ocotber 26th.—Subsequent to my last visit the patient was transferred by Dr. Salzer to the care of Dr. Simon, who tells me that uræmic symptoms developed about the 23d and the patient died comatose on the 25th. There was no autopsy.

CASE III. *Dilatation of the Stomach; Tumors in Epigastric and Right Hypochondriac Regions.*—A. P., aged forty-seven years, seen October 19th with Dr. Jarrett, of Towson, complaining of indigestion and stomach trouble. His personal and family history are excellent, though he states that one brother died of a tumor in the abdomen. For the past six months he has been in failing health and has had distress after eating, usually within half an hour, sometimes as soon as ten minutes. The vomitus consists of the food he has taken, never any blood. He has never vomited any very large quantities. The pain is marked after eating and becomes more severe until the contents of the stomach are ejected. On several occasions he has passed blood in the stools, but he thinks this comes from hæmorrhoids which he has had for many years.

The patient is a very well built man, looks thin, but is a little sallow, scarcely cachectic. Tongue is red, clean, and indented.

Abdomen.—Walls thin. Occupying the left epigastric region there is a large projection which varies in shape and in prominence. Definite peristalsis is to be seen, but the waves do not pass beyond the middle line. This bulging during peristalsis occupies the left epigastric and the upper right quadrant of the umbilical regions.

On palpation, the abdomen is everywhere soft, very resistant just below the ensiform cartilage and over the prominence above noted. In the latter the resistance varies with the presence or absence of the peristaltic waves. Immediately below the ensiform cartilage there is a definite ridge-like swelling which is superficial, very tender, and does not extend entirely across the space between the costal margins. It has a boardy hardness. On drawing a deep breath the fingers can be placed directly above it and it descends about an inch. In the left lumbar region, just below the tenth rib and the adjacent costal margin, there is to be felt a firm mass, extending seven centimetres in a vertical direction. Anteriorly in reality it can be felt within the right epigastric region, and outward it extends to nearly the mid-axillary line. On deep inspiration it descends and gives one some-

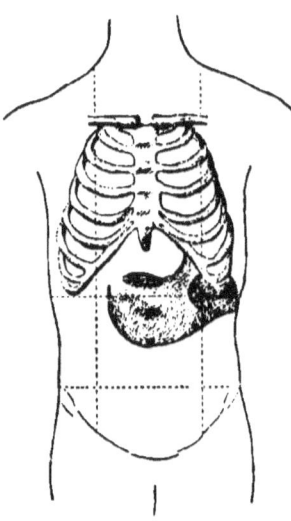

FIG. 4.—Showing the position of the tumor masses in Case III.

what the impression of a rounded body, and on bimanual palpation it is not very movable.

The edge of the spleen is not palpable ; neither kidney can be felt ; the edge of the liver is not palpable ; nor does there appear to be any definite enlargement of the organ. On inflating the stomach the prominence in the epigastric and umbilical regions becomes very marked and its lower curve extends to a little below the navel. The upper limit of stomach tympany is just at the sixth rib in the nipple line. There are no glandular enlargements. The patient became gradually worse and died about Christmas time.

CASE IV. *Dilated Stomach, forming a Visible Tumor; an Oblong Mass in the Right Epigastric and Umbilical Regions.*—Annie D., aged forty-eight years, Bohemian, admitted October 1st, complaining of swelling in the abdomen, pain in the back, and vomiting.

She knows of no hereditary disease in her family. Her husband died of tuberculosis.

Patient was always strong and well ; she has had three children. Her present trouble began eight months ago with pain of a dull, aching character in the stomach, and dyspepsia, but until recently she has had no vomiting, and has kept about and at work up to a week ago, when she began to vomit. Prior to this she noticed that the abdomen was swollen. The vomiting has been chiefly after taking food, and she has not brought up any large quantities.

Present Condition.—Patient is thin, but the emaciation is not extreme. The lips and mucous membranes are of ·a fairly good color. Tongue is slightly furred with a white coating. Pulse regular ; temperature normal ; superficial glands not enlarged.

The abdomen is prominent, particularly in the umbilical and left hypochondriac regions. Under observation there occur in these parts undulatory waves of peristalsis, and the outlines of the stomach become unusually distinct, the greater curvature reaching fully three inches below the level of the navel, the lesser curvature just above this point. As the waves of contraction pass there is a vertical constriction just to the left of the middle line. The peristalsis comes on spontaneously, and any stimulus, such as flipping with a towel or even palpation, at once excites it. On palpation, except during the time of the contraction referred to, the abdomen is everywhere soft. Just above and to the right of the navel there is to be felt an oblong mass, which takes a direction upward and outward toward the costal margin. It is oblong, slightly movable, firm, smooth, and not painful.

On October 3d a test breakfast was given at 8 A. M. At nine o'clock the stomach tube was introduced and about a quart and a half of very sour-smelling, brownish material removed. During the passage of the stomach pump the patient vomited, and she felt very faint. The examination for free hydrochloric acid was negative.

The patient left the hospital on October 14th in much the same condition, and has not been heard of since.

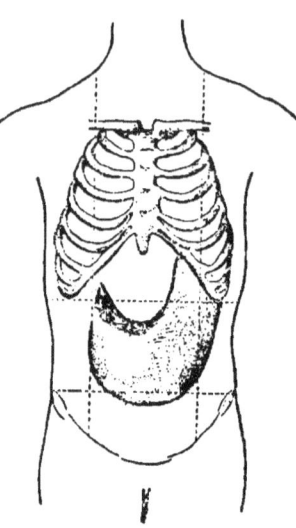

FIG. 5.—Showing the position of the tumor and the outlines of the stomach in Case IV.

CASE V. *Remarkably Movable Tumor of Pylorus; Dilated Stomach; Gastro-enterostomy.*—Mary M., aged fifty-eight years, colored, admitted on October 26th, complaining of pain in the abdomen and vomiting.

She has been a healthy woman, married twelve years; has had six children and four miscarriages. She has always had very good health up to the onset of present illness, which began in June with burning feelings in the chest and pain after eating, sometimes vomiting. These symptoms have continued with variations. At times she would be better, and then she would have spells of belching and vomiting. She had often vomited large quantities of liquid. She makes no complaint except of the stomach symptoms. Lately she has been very constipated.

Abdomen.—The walls are very loose, flabby, thrown into many folds. In the right hypochondriac and right epigastric regions there is a marked rounded prominence, which below extends to within two centimetres of the navel, and reaches nearly to the middle line. It descends slightly with inspiration. On palpation, this proves to be a solid mass, which can be grasped and is freely movable. It is irregular, rounded, not reniform, but is smooth at its upper and right borders, more irregular below and to the left, but a definite hilum is not to be felt. To the touch there is conveyed a sense of firm yet elastic resistance, such as is given by a solid organ. On prolonged palpation no gas is felt passing through it. It is extraordinarily mobile, and can be pushed into the epigastric region far over into the right hypochondriac region, and below into the right lumbar and umbilical regions to a level with the line of the anterior superior spines. On firm pressure, the lower margin can even be forced into the iliac region. The diagram, which was made with great care, illustrates the various positions which the mass can be made to assume. It can also be pushed into the right hypochondriac region, so as to *be covered almost completely by the ribs*, and in subsequent examinations this was not infrequently the situation in which it was found, and from which it could only be dislocated by the deepest inspiration or by deep pressure in the renal region. The mass is not tender even on firm pressure. There is dullness over it, but not complete flatness. The patient notices that the mass

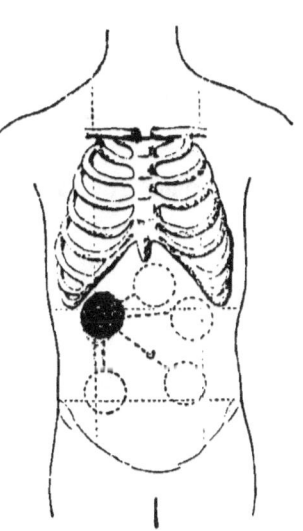

FIG. 6.—Showing the positions into which the tumor could be placed in case V.

changes in position as she moves about, and when she sits up it moves far down into the abdomen, while when on her back it is frequently beneath the right ribs. When this mass is out from beneath the right costal margin the right kidney can not be felt, nor on the left side, on the deepest inspiration, could the kidney be palpated. Behind there are depressions in the renal regions.

The edge of the liver can not be felt ; the area of splenic dullness is not increased ; the edge can not be reached even on deep inspiration.

A test breakfast, withdrawn an hour and ten minutes after, gave two hundred cubic centimetres of fluid in the stomach, which contained no free hydrochloric acid. The stomach was inflated with gas, and the outline of the greater curvature reached almost to the navel. When the gas was in the stomach, palpation of the most careful character gave no sensation of any fluid passing through the tumor.

This patient came in with a diagnosis of probable cancer in the stomach, which the history of repeated attacks of vomiting and progressive loss of weight and the existence of a tumor in the abdomen seemed to justify. Extreme mobility is a feature of certain tumors of the pylorus, as in the specimen which I showed at the Philadelphia Pathological Society of solid tumor of the pylorus, about the size of the mass under consideration, which could be moved readily into either hypochondriac region, and which was sometimes completely under the ribs and out of reach. The autopsy showed it to be a tumor of the pylorus. The possibility of such cases has to be considered in speaking of the nature of the present one. Here the mass is of unusual mobility, and can be passed into the renal region on the right side. It has not a reniform shape, but it has the consistence and the resistance of the kidney. A point very much in favor of its renal character is the mobility downward, and the tumor of this sort which can be pushed up beneath the ribs and also far down to the iliac regions is certainly highly suggestive of floating kidney. Another important fact is that, in a woman with such a lax abdominal wall, no right kidney can be felt. The gastric disturbance and dilatation of the stomach present are both explicable on the view that this tumor mass has compressed the duodenum and caused secondary dilatation. Nor is this, considering the history of so many cases, inconsistent with the view that the tumor mass may be really a kidney. On the other hand, the tumor has not the shape of a kidney, and a distinct hilum can not be felt. No left kidney can be palpated, and it may be that this is an instance of conglomerate kidney, such as was found in Polk's celebrated case.

At any rate, I have suggested to Dr. Halsted that an exploratory laparotomy be made, and if it is found to be a movable kidney, the organ can be stitched into position.

November 4th.—The patient has been better for the past few days. She has had her stomach washed out early in the morning. To-day at ward class a careful examination was again made. The tumor mass was evident just beneath the right costal margin, and it was difficult to displace it from this point by the deepest inspiration; but, on turning on the left side, it readily fell over toward the umbilicus, and had practically the mobility noted before. The stomach was again inflated, and the outlines became remarkably plain. The greater curvature was just about the level of the navel, somewhat above the level previously noted. The peristalsis was unusually distinct.

5th.—This morning at 10.30 Dr. Halsted operated, making a long vertical incision over the right rectus. When the peritonæum was opened the tumor mass was directly exposed, and found to be a solid growth of the anterior wall and lesser curvature of the stomach in the pyloric region. There were no adhesions; the stomach was much dilated. He at first intended to resect the tumor, but, on examining the retro-peritoneal glands, they were found to be enlarged, and it was thought best to do a gastro-enterostomy. The patient died two days afterward.

CASE VI. *Tumor in Left Epigastric Region; Dilatation of the Stomach.*—In consultation with Dr. Barclay I saw to-day, December 6th, A. B., aged sixty-four years, a German.

Patient had been in failing health for some months, and had had dyspepsia for several years. He, however, kept about and at his work until early in October, when he consulted Dr. Barclay for jaundice, which seems to have been intense and to have come on suddenly, not, however, with much pain and not in a way suggestive of gallstones. He had never had jaundice before, but had one or two attacks of pain resembling that of gallstone colic. On examination, the left lobe of the liver was found to be enlarged and a tumor mass occupied the whole of the epigastric region. It was tender, not fluctuating, and the doctor regarded it as an enlarged left lobe of the liver. He had moderate fever. After the persistence of these symptoms for some weeks he vomited a quantity of pus, the tumor mass gradually disappeared, and the jaundice became less intense. The gastric symptoms, however, continued and he began at intervals to vomit large quantities of dark-brown material, containing undigested remnants of food. The doctor has washed out the stomach with great relief, but he has gradually failed and has become more anæmic.

Present Condition.—The patient is fairly well nourished; face is not especially emaciated, and he has not a cachectic look. There is no jaundice. The temperature is normal; pulse about 96, of fairly good volume; tongue is slightly furred.

Abdomen.—Panniculus is well preserved. The upper zone is prominent, particularly in the left hypochondriac region, and at intervals a distinct hemispherical prominence appears below the left costal margin, and waves of peristalsis are seen passing from left to right. The prominence is noticeable as far as the navel, but a definite contour of the stomach is not visible. Midway between the ensiform cartilage and the navel and a little to the left there is a tumor-like prominence which moves with the descent of the diaphragm. On palpation, the abdomen is everywhere soft, quite painless on pressure, and the tumor mass just described is felt as a firm, solid body about five centimetres in vertical extent and about six centimetres in transverse extent. It is entirely to the left of the middle line. It is firm, smooth, not painful, except on very firm pressure, and is not movable. It descends about four centimetres during inspiration. No gurgling is felt in it. Nothing is to be felt to the left of the median line in the pyloric region. Splashing can be readily obtained, and on percussion the stomach tympany extends to a finger's breadth above the navel. The upper limit of the liver dullness in the mammillary line is at the seventh rib, and it does not extend beyond the costal region. The left lobe of the liver is not palpable; the dullness is at the juncture of the fifth costal cartilage with the sternum, and extends three fingers' breadth. It can be separated from the dullness over the tumor mass in the epigastric region. The spleen is not enlarged. There are no superficial glandular enlargements.

The patient had had a severe attack of vomiting this morning, and his stomach was not nearly so much dilated as usual. The vomited matter which I saw had the usual characters—dark brown, with frothy scum. The urine was somewhat diminished, and he complained very much of thirst.

Two points of interest present themselves in this case, which otherwise seems to have all the characters of ordinary dilatation of the stomach from pyloric obstruction. In the first place, the nature of the attack of severe jaundice with the tumor mass in the epigastric region. From Dr. Barclay's account, there can be no question that the patient vomited a large quantity of pus, and that subsequent to this the tumor disappeared and the jaundice got better. There are two suggestions in this connection: that there was a large carcinoma of the stomach, with suppuration at its base and about the tissues of the gastro-hepatic omentum, with compression of the bile ducts. Suppuration does occur at the base of malignant growths, more particularly when they form adhesions with adjacent organs, and I have placed such instances on record; indeed, there may be a considerable collection of pus between the left lobe of the liver and the stomach. The other suggestion is that the jaundice and enlargement of the left lobe of the liver were associated with gallstones and suppuration in the region of the ducts, with discharge into the stomach, and subsequent cicatricial contraction about the pylorus and dilatatio ventriculi. The jaundice, however, would scarcely have disappeared, and this is not a very likely condition. And, lastly, it is interesting to note here the situation of the tumor mass—not at all in the position usually felt in carcinoma of the pylorus, but, as will have been noted in the histories of the other cases, the tumor is extremely variable in position. Though in a somewhat unusual situation, it is quite possible that this really may be a tumor mass causing the stenosis of the pyloric orifice.

Dr. Barclay writes me that this patient died on the 29th of December of exhaustion. The post-mortem showed an enormously distended stomach, which covered the intestines like an apron. The pyloric orifice was surrounded by a large mass of cancer, which so nearly occluded it that only the tip of the finger could enter. The cancer extended also slightly into the duodenum and on the posterior wall of the stomach, which showed extensive ulceration. The left lobe of the liver was shrunken and showed soft, nodular masses breaking down into pus. The mesenteric glands were enlarged and cancerous. The bile passages and gall bladder were normal.

The post-mortem does not throw much light upon the early history of this case, the symptoms of which came on apparently with jaundice and enlargement of the left lobe of the liver. The tumor mass proved to be, as was supposed, at the pylorus, though in a somewhat unusual situation—entirely, at the time of my visit, to the left of the middle line.

CASE VII. *Dilated Stomach; Tumor at the Pyloric Orifice.*— *October 1st.* I saw to-day, with Dr. W. B. Perry, Mrs. R., aged about sixty years, complaining of dyspepsia and vomiting. She had been a healthy woman until about a year ago, when she began to have attacks of dyspepsia and occasionally of vomiting. These symptoms have become progressively aggravated and she has within the past three months lost flesh rapidly. A marked feature in the case has been the vomiting at intervals of very large quantities of a brownish liquid mixed with portions of food.

Patient is a small-framed woman, much emaciated, and looks very feeble. The abdomen is greatly distended, chiefly on the left side and below the level of the navel. The nature of the trouble is at once apparent by the active waves of peristalsis which, as they pass from left to right, bring out with unusual distinctness the contour of the greater and lesser curvatures, the former passing at a level of about three inches above the pubes, and the latter midway between the navel and ensiform cartilage. The organ becomes unusually hard and firm. Far over to the right, just at the border of the epigastric and umbilical regions, there is to be plainly felt an irregular, nodular mass, which is movable and is depressed on inspiration. No gas is felt passing through it, but the position and characters suggest a pyloric cancer. Lavage had been already practiced for some time, but she was in too feeble a condition to expect much from any treatment.

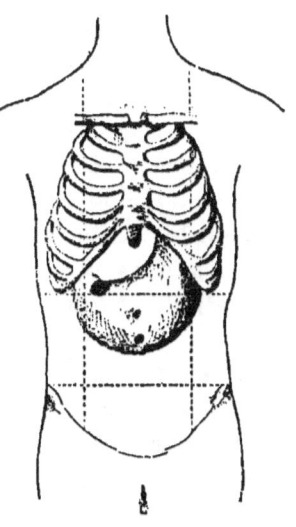

FIG. 7.—The position of the nodule and outline of the stomach in Case VII.

Dr. Perry writes that the patient died on October 4th.

CASE VIII. *Dilatation of the Stomach, forming a Visible Tumor; Nodular Mass at the Pylorus.*—January 10, 1893, Captain ——, of Virginia, patient of Dr. R. J. Hicks and Dr. Salzer, came complaining of dyspepsia and discomfort after eating.

The patient has been a healthy man, a free liver, and a late sitter; irregular in his meals. He has not been a chronic dyspeptic, and has only had an occasional attack of indigestion until the onset of the present trouble. From Christmas, 1891, he has been ailing, though able to attend to his business. He has had loss of appetite; no special nausea, and has never vomited any large quantity. After eating, however, particularly an ordinary meal, he has feelings of uneasiness and distress, and rumbling and distention in the upper part of the abdomen. He has never had any severe pain, but a sense of uneasiness when the stomach is full and occasionally a griping pain. Ever since the attack of diarrhœa following the influenza he has had obstinate constipation. There has been persistent loss of weight, from a hundred and ninety to a hundred and forty-five pounds. Though attending to his business, he is at times very weak and feeble, and feels that he has lost a great part of his former vigor.

Present Condition.—Thin, not extremely emaciated, not cachectic; color of mucous membrane good. Abdomen a little below level of costal margin; marked fullness in epigastric and umbilical regions, leaving a definite furrow along right costal margin. During observation distention becomes much more marked, and at intervals the outline of the stomach is unusually distinct. Waves of peristalsis pass actively from left to right, and the lower limit of the stomach is seen to be at least a finger's breadth below the navel. To the right it extends almost to the costal margin opposite the tenth rib. The peristalsis is unusually active, waves passing every few moments, and during their passage the stomach walls become hard. Gas can be heard bubbling through the pylorus.

Palpation.—Everywhere soft; no special resistance except over the stomach itself when in contraction. The pylorus can be felt in the parasternal line at a point midway between the navel and the tip of the tenth costal cartilage. Here is a firm thickening about the size of a large walnut. Though this is a little far out and low for the situation of the pylorus, yet the stomach is a good deal depressed and the whole pyloric pouch lies to the left of the middle line. There is nothing special to be felt along the line of the lesser curvature. There is a little resistance between the costal margin and the navel, which is probably due to the right lobe of the liver. Gas is not felt to bubble through this pyloric mass, nor does it seem to vary in resistance and hardness.

January 22d.—Patient came into the private ward under my care, chiefly to determine whether an operation, which had been suggested by Dr. Salzer, was advisable or not. On admission, the stomach was very much in the condition mentioned in the previous note. Ewald's test breakfast, withdrawn an hour after, yielded two hundred and fifty cubic centimetres of a clear, slightly yellow fluid containing partially digested bread. The odor was sour; the tests for free hydrochloric acid were negative.

He had at times a great deal of distress, owing to the active character of the peristaltic movements. He was placed on a diet of milk, beef juice, and egg albumin, small quantities being given every two hours. The stomach was thoroughly emptied night and morning.

Within a few days this treatment made the greatest change in the condition of dilatation; the organ reduced greatly in size, the waves of dilatation were no longer evident, and he felt much more comfortable. The reduction in the dilatation made a very marked change in the tumor mass above mentioned. Instead of a small, nodular body to be felt far over to the right, there was now evident to the right of the parasternal line, in the epigastric region, a large, solid mass of the size of an egg. In spite of the improvement in the local condition, his general strength failed with rapidity, and on the 28th it was thought advisable for him to be removed to his home, where he died early in March.

CASE IX. *Nodular Tumor Mass in the Region of the Pylorus; Dilatation of the Stomach.*—Rachel C., aged sixty two years, admitted February 13th, complaining of a lump in the right side of the abdomen. Mother died of pulmonary tuberculosis.

With the exception of pulmonary hæmorrhages, of which she has had in all nine attacks since her nineteenth year, and an attack of transient left-sided hemiplegia of five days' duration, she has been a healthy woman. Has had ten children. No special history of dyspepsia.

On September 15th she had for the first time severe pain in the upper part of the abdomen, which continued for nine days, and was intense enough to keep her from sleep. Soon after this she felt a lump in the right side, which has all along been painful and associated with a dragging feeling when she lies on the left side. At intervals she has a puffed, distended feeling in the abdomen with diffuse soreness. Her appetite has been poor, and she has had nausea sometimes after eating, and has several times at night had attacks of vomiting. She has had no swelling of the legs. The urine is clear; no special diminution in amount. Patient is emaciated, sallow; tongue coated, white; pulse is 88, regular, tension a little increased. There is a soft systolic murmur at the apex; with the exception of hyper-resonance in the front and sides, and exaggeration and prolongation of the expiratory murmur, there are no changes to be noted in the lungs.

Abdomen.—There is a slight prominence in the right hypochondrium, and on inspiration a tumor mass can be seen to descend in the parasternal line. On palpation in this region, just below the costal margin, there is a hard, rounded mass, the outline of which can be pretty clearly made out toward the median line, but toward the right it is apparently continuous with the margin of the liver. It is superficial, nodular, hard, and very painful. On deep inspiration, it descends almost to the level of the navel, and the fingers can be then placed between the tumor and the liver margin, and it can be held down. Gas can be felt bubbling through the mass. On percussion, there is a flat tympany. When quiet and in repose no peristalsis can be seen as a rule. When the patient turns over on the left side the mass falls forward and to the left, and can readily be grasped between the hands. The epigastric region is a little flattened; sometimes distinctly depressed. The lower umbilical region, on the contrary, is full.

On dilatation with tartaric acid and bicarbonate of sodium, the stomach is seen to be depressed and dilated. The lesser curvature passed just at the level of the navel; the greater curvature at a distance of seven centimetres below. Waves of peristalsis were then seen in the stomach walls passing from left to right, and sometimes the organ showed an hour-glass contraction.

The test breakfast showed the presence of the organic acids, absence of free hydrochloric. Material washed out was dark in color and smelled very sour.

The liver is not enlarged; the spleen not palpable; no enlargement of the external glands.

CASE X. *Enormous Dilatation of the Stomach, forming a Visible Tumor; Ridge-like Thickening in the Pyloric Region.*— Emma H., aged thirty years, colored, admitted (to the gynæcological ward and transferred) April 28, 1893, complaining of swelling of the abdomen, nausea, and vomiting. Nothing of any moment in the family history. She was healthy until about eight months ago, when she began to have dyspepsia and distress after eating. She has not infrequently had attacks of vomiting. Lately she has been much nauseated after eating, and has had pain and swelling of the abdomen, with much belching. Six months ago she brought up food mixed with blood. Lately she has only been vomiting food. She has lost in weight within the past year, and has noticed that she passes much less urine than formerly. The patient is moderately wasted; weight, a hundred and five pounds; lips and mucous membranes of good color. Temperature normal; pulse, 112. The abdomen is flaccid, but a little prominent, and, on inspection, very marked waves of peristalsis are seen passing from left to right. They occupy a considerable area, extending from just below the costal margin to midway between the navel and the pubes. As they pass, the skin is lifted in very definite prominences. On palpation, there is very marked succussion, and, on changing from side to side, the dullness alters as the fluid sags with the change of position. On inflating the stomach, it is found to occupy nearly the whole abdomen. The tympany begins above at the fifth rib and extends to the pubes. The lesser curvature is seen just above the umbilicus. There is very prominent distention in the pyloric region, and the gastric tympany extends nearly to the right anterior superior spine. No nodular masses or tumor could be felt. With the stomach tube a large quantity of a greenish-yellow liquid with remnants of food was removed. When the stomach was emptied there then could be felt midway between the umbilicus and the costal margin a ridge-like mass about the size of the thumb, which was freely movable and descended with inspiration, and which subsequent examinations showed was extremely variable. not being palpable when the organ was very greatly distended. With lavage and feeding at short intervals the patient improved very much and the stomach reduced very materially in size and she gained in weight.

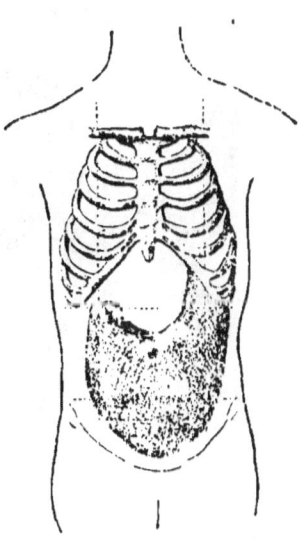

Fig. 8.—Outline of the stomach in Case X, showing the position of the ridge-like mass.

From the prolonged history of dyspepsia and the fact that she had on several occasions vomited blood, this was very probably an instance of dilatation from the cicatrization of an ulcer ; and the small, elongated nodular thickening in the region of the pylorus also suggested this condition.

General Remarks on the Diagnosis of the Tumor caused by Dilated Stomach.—Inspection gives most important information, the value of which may be gathered from the fact that in these ten cases the diagnosis was made *de visu*. Bear in mind, in the first place, that a dilated stomach may occupy every region of the abdomen except the upper part of the epigastric and may form a very prominent tumor.

FIG. 9.—Profile view of the abdomen of Sarah A., aged sixty-five, showing the tumor formed by the dilated stomach. From a photograph taken during life.

The photographs (Figs. 9 and 10) which I show you illustrate this very well. They were taken during life from a woman, aged sixty-five years, who was admitted to the hospital complaining of attacks of vomiting which had persisted for nearly two years, during which time she had become gradually emaciated and very weak. She had at intervals brought up enormous quantities of fluid. On inspection, the abdomen was greatly distended, particularly on the left side and in the umbilical and hypochondriac regions. It was uniform, but at intervals there were slight irregularities and elevations; no marked waves of contraction. On palpation, the abdomen was everywhere soft, except at a point to the right of and just below the navel, where there was a hard, resistant mass. At first it seemed scarcely possible that the entire abdominal distention could be due to a dilated stomach, but the reduction in size after

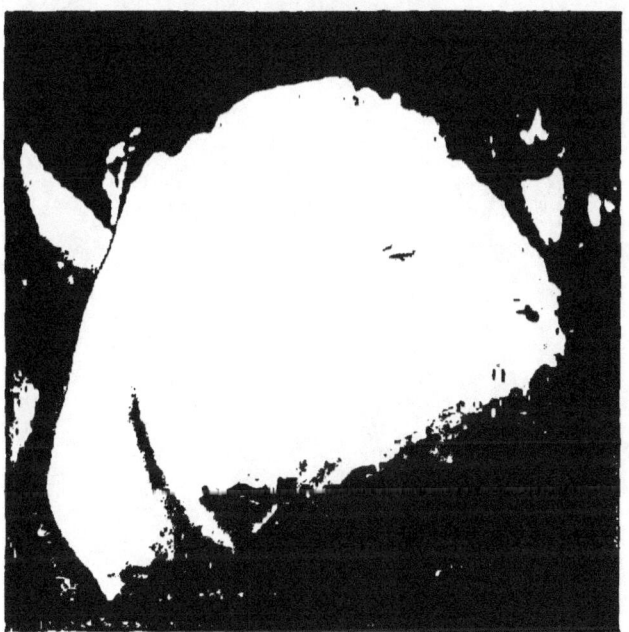

FIG. 10.—Tumor of the abdomen caused by a dilated stomach; case of Sarah A. From a photograph taken during life.

vomiting and after lavage, the depth to which the stomach tube could pass, and the irregular waves of protrusion left no doubt that the distention was due to an enormously dilated stomach. She died November 16, 1889. The photographs (Figs. 9 and 10) show the profile and front views taken during life, and Fig. 11, from a photograph taken after death, shows the position of the organ and its enormous enlargement. There was cancerous stricture of the pylorus.

The most prominent distention is usually in the left

half of the umbilical region, but it may be chiefly below the navel. A definite stomach contour may be seen very plainly in many instances of dilatation from stenosis of the pylorus. At intervals, during the contraction of the

Fig 11.—Showing the position and size of the stomach in Sarah A. From a photograph taken at the autopsy.

stomach walls, the outline of the greater curvature descends on the left side, curving at a level of the anterior superior spine, and passing to the right at a variable distance above the pubes, sometimes not more than three or four centimetres, sometimes midway between the pubes and the naval. Curving upward, it ends either in the left lumbar or more frequently in the right upper quadrant of the umbilical region, sometimes appearing to pass beneath the right costal margin. The lesser curve is frequently much more distinct, the line passing vertically parallel with the left border of the sternum or in the parasternal line, curving to the left of the navel, and often during the contraction of the organ forming a very well marked, sharply defined contour at or a little below the level of the navel. I have found the greatest surprise expressed by practitioners that the stomach should be so low, that even the lesser curvature should be below the level of the navel; but this is frequently the case in extreme dilatation. In the first place, then, the outline of the organ may give to you at a glance the diagnosis. Secondly, inspection is of the greatest value in determining the presence of peristalsis. Though enormously stretched, there is hypertrophy of the muscular coats and great increase in the activity and frequency of the movements. In all of the cases they were present. The movements are of two kinds, which may be seen together or separately: First, peristaltic waves, passing slowly from left to right, more rarely antiperistalsis, from right to left. The mere exposure of the abdomen to the cool air is usually sufficient to excite them. Sometimes the stimulus of palpation is required, or the flapping of the skin with a wet towel. During the passage of these waves the outline of the organ becomes evident; sometimes, as already noted, the greater and lesser curvatures are plainly to be seen. Sometimes, too, as the waves reach the pyloric region a tumor mass may be rendered visible or made more prominent. The stomach may be so enormously dilated that the walls are in a condition of paralytic distention and no peristaltic waves are seen, as in the case from which Figs. 9 to 11 were taken.

A second variety of movement to be seen in a dilated stomach consists in a slowly performed irregular protrusion here and there of the stomach wall, which lifts the skin of the abdomen in a hemispherical boss or prominence. This may develop at any point, more frequently toward the greater curvature. They usually occur with the peristaltic waves and in combination may throw out in bold relief the contour of the organ, sometimes also giving to it a somewhat hour-glass shape, owing to corresponding depressions about the middle of the greater and of the lesser curvatures. The upper depression is seen in Fig. 1. These irregular protrusions may be seen in enormously dilated stomachs, in which the peristaltic waves are no longer visible, as in the case just mentioned, of which I have shown you the photographs. Let me again emphasize the value of inspection by reminding you that of the thirteen instances, in ten the diagnosis was manifest on inspection alone.

Palpation.—Four points may be determined by this procedure. The existence of the splashing sound or succussion, the *clapotage*, which is always present, and which, though in no way diagnostic, yet is of value in connection with a prominence in the left flank and lower umbilical region. It is of use also in determining the lowest level of the organ. With the hand on the abdomen, as the peristaltic waves pass, or as the irregular protrusions develop, you will notice that the organ hardens; and toward the pylorus, as the wave is followed, there may be an exceedingly firm contraction. After persisting for a minute or so the muscular walls relax and are again soft and readily depressed. In some instances the muscular contraction at the pylorus is extremely firm and hard, and the relaxation beneath the hand reminds one of that of the uterus. A third point of importance, particularly in palpation of the pyloric region, is the gurgling of gas through the pyloric orifice. This is usually very marked when the stomach is inflated, but it may occur spontaneously and in some instances at regular intervals. In doubtful tumors of this region this is a sign to which scarcely sufficient attention has been paid. Its importance will be referred to again in connection with tumors of the intestines, as in one case in the series it saved us from a somewhat serious error. And lastly it is by palpation chiefly that we are enabled to determine the presence or absence of a pyloric tumor. And here you have to bear in mind that in dilatation of the stomach the pyloric tumor may be extremely variable, readily felt to-day, scarcely palpable to-morrow, dependent very much upon the grade of distention. You will find this very strikingly illustrated after washing out the stomach, when perhaps a comparatively small pyloric mass may be found to be quite large and prominent. When the organ is extremely dilated, the tumor may be scarcely palpable. This was particularly well illustrated in Case VII, which was sent to me by Dr. Salzer for an opinion as to the advisability of a Loretta's operation. The tumor at the pylorus, which at the first examination seemed not larger than a walnut, after thoroughly emptying the stomach was found to be a solid mass the size of an egg.

Percussion combined with palpation brings out most clearly the splashing sound, which in cases of extreme dilatation may be most evident below the transverse navel line. The extent of stomach tympany will vary with the position of the patient. In the recumbent posture it may extend in the nipple line from the fifth costal cartilage to within two or three fingers' breadth of the pubes. In the erect posture a line of transverse dullness may be accurately defined, which will sink as the patient is gradually placed in the recumbent position. The extent of stomach tympany varies, of course, with the amount of fluid contents, and after the attacks of vomiting in which large quantities of liquid are brought up it may be very much extended.

In doubtful cases inflation of the organ is a most valuable method, and is best accomplished by the use of the bicarbonate of sodium and tartaric acid, from half a teaspoonful to a teaspoonful dissolved separately and taken one after the other, the patient being directed to refrain, as far as possible, from belching. Inspection may, through thin abdominal walls, at once reveal the distended stomach, displaying active peristaltic movements. The percussion limits can then be also more definitely defined. Palpation in the pyloric region may give the sensation of gas bubbling through into the duodenum. This method of inflation is more satisfactory on the whole than that of pumping air into the stomach. When gastric ulcer is suspected these proceedings should be practiced with great caution or omitted altogether.

Auscultation gives little information of value. One hears the sizzling sound of the gas as the contents of the stomach are churned about; sometimes quite loud gurgling is heard as the fluid passes through the pylorus, often loud enough to be heard at a distance. The succession splash may be obtained by placing the ear upon the abdomen, and either shaking the patient or asking him to depress suddenly the diaphragm.

The characters of the contents of the dilated stomach, the general symptoms, and special features I shall not discuss, as the subject before us is more particularly the form of abdominal tumor caused by it. To one special point, however, I may refer, as it is of some moment in the treatment of these cases of dilatation from stenosis of the pylorus. The recent experiments of von Mering show that water is not absorbed to any extent from the stomach, but is passed into the intestine, usually at regular intervals, by a rhythmical opening and closing of the pylorus.

Not only is the resorption of water extremely slight, but hand in hand with the absorption of the sugars and peptones there is actually a secretion of water corresponding in some measure to the amount of substances absorbed. With these facts correspond closely certain points in the history of dilatation of the stomach. The organ is never empty, and even after it is pumped out as much as possible, fluid will reaccumulate without any liquid having been taken; and frequently patients will remark in astonishment that the amount which they have vomited far exceeds in quantity what has been taken by the mouth. This explains also other striking symptoms in excessive dilatation of the stomach—namely, the great reduction in the amount of urine secreted, the dryness of the skin, and the wasting, which is proportionate to the degree and persistency of the dilatation rather than to the nature of the obstruction. Unverricht has suggested to supplement this water depletion by the use of large enemata, two litres daily of salt solution, the use of which he states has been followed by marked improvement in the symptoms, and in some instances by an increase in weight.

And, lastly, there is the question, What conditions may be confounded with dilatation of the stomach? Nothing, in reality, if the examination is made systematically and thoroughly. The physical signs alone are generally sufficient, and, when taken in connection with the general symptoms, quite distinctive; thus there was not one of the cases in this series about which a shade of doubt existed. Yet mistakes have arisen, some of a remarkable nature, owing to ignorance of the fact that the dilated organ may be chiefly to the left of and below the umbilicus. The tumor has been mistaken for an ovarian cyst (*Detroit Lancet*, January, 1880), and even after tapping and the withdrawal of a dark-colored fluid containing grains of rice, pieces of potato, bread, meat, etc., laparotomy was performed for ovarian tumor. The enormous dilatation, as shown in Figs. 9 and 10, with paralytic distention and absence of the peristaltic waves, might, and indeed did, when the stomach was very full, simulate ascites, but no serious difficulty could arise in the differentiation. Tumors of the colon, causing obstruction, lead to great distention of the large bowel, in which active waves of peristalsis may be seen passing from right to left. Usually the abdomen is distended more uniformly or chiefly in the epigastric zone, and intestinal, not gastric, symptoms have been present, and the inflation of the stomach alone, or, if practicable, combined inflation of the stomach and colon, will usually give information upon which you may base a definite conclusion.

A dilated stomach, causing a prominent tumor of the abdomen, is almost invariably due to stenosis of the pylorus. As already mentioned, in all of the cases a tumor was evident, and in all the condition was that to which the term dilatation of the stomach is more correctly limited. In rare instances a prominent tumor may be caused by muscular insufficiency, as it is called, or atony of the stomach, and occasionally, by change in the position of the organ, the so-called *descensus ventriculi*. The differential diagnosis of these conditions you will find fully given in the special works above mentioned. I may, however, remark that only in very exceptional instances of atony of the stomach or of *descensus ventriculi* are the peristaltic waves seen. In women who have borne many children, and who have the extremely relaxed abdominal walls, the condition which Glénard has termed enteroptosis may be associated with great depression and enlargement of the stomach. In some cases the decision is very difficult. I show you here the stomach outlines of a patient who at first we thought had dilatation of the stomach from pyloric obstruction. The organ reached nearly to the pubes; the lower curvature was, however, at the umbilicus. The vertical measurements of the stomach were twenty centimetres in the middle line, thirty-one centimetres from the lower border of the eighth costal cartilage to the middle of Poupart's ligament, and the transverse diameter twenty-eight centimetres and a half. Occasionally there were to be seen peristaltic waves crossing from left to right. No tumor was at any time to be felt. The liver was depressed; the kidneys were movable. The examination for free hydrochloric acid was variable; sometimes it was present, sometimes absent. She had had dyspepsia for some years, and within the year much belching and some vomiting, but never of very large quantities of liquid. With lavage and careful dieting she improved very much, and gained fourteen pounds in three months. We subsequently lost sight of her.

II. TUMOR FORMED BY CONTRACTED STOMACH.—There are two conditions in which the stomach itself, in a state of contraction, may form a definite, palpable tumor—first, in occlusion of the œsophagus, when the organ shrinks and may be sometimes felt as a narrow, firm cord, lying below the margin of the left lobe of the liver; and, second, when there is diffuse thickening of the stomach walls with contraction of the lumen in cirrhosis or in diffuse cancerous infiltration. The following is a remarkable instance of the latter condition, which presented a well-marked tumor:

CASE XI. *Tumor in the Epigastric Region, consisting of the Stomach diffusely Infiltrated with Carcinoma.*—George H., aged sixty years, tailor by occupation, German, admitted on April 4, 1893, complaining of indigestion.

Has always been strong and well; had gonorrhœa when twenty; has been a moderate drinker; does not use tobacco.

Present illness began about five months ago with uncomfortable feelings in the epigastrium and constipation. Prior to this he states that he had no dyspepsia. Soon he began to have much distention after meals, and water brash. He has never had any vomiting, but he spits up a great deal of mucus. There is sometimes a sharp pain in the abdomen, but, as a rule, there is only a heavy, uneasy sensation after eating. In January he noticed that there was a swelling above the left clavicle. In March he had one or two attacks of vomiting, and he has been much troubled with hiccough.

He applied at the dispensary about the end of January, and a stomach tube was introduced, but it brought up a little blood. Since his illness began he has lost forty-five pounds in weight.

Present Condition.—Considerable emaciation; lips and mucous membranes of good color. Above the left clavicle the lymph glands are enlarged and hard; slightly enlarged above the right. They are also somewhat enlarged in the inguinal region. The thoracic organs are normal.

The abdomen is flaccid, symmetrical; a little full, perhaps, in the epigastric region. On palpation, a ridge-like mass is to be felt in the left hypochondrium, which extends across the middle line to the right side as far as the parasternal line. It descends with deep inspiration as low as the umbilicus. The lower edge is very distinct and feels somewhat like a rolled omentum. It can scarcely be separated above from the liver, the edge of which is just palpable. There is a flat tympany over the mass. On inflation of the stomach, vomiting occurred and a good deal of distress.

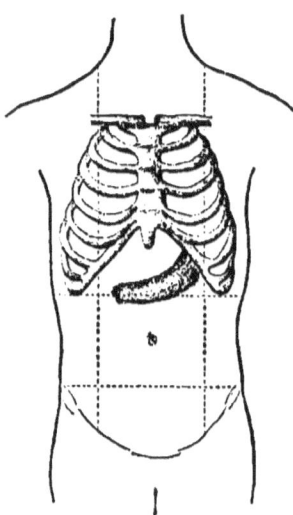

Fig. 12.—Showing the position of the tumor mass consisting of diffusely infiltrated stomach walls in Case XI.

The urine was negative.

On April 21st the patient had tenderness along the saphenous vein and the calf of the leg, and there was œdema of the ankle and of the leg.

The patient left hospital on May 5th, and died two days afterward at his home, where the autopsy was made by Dr. F. R. Smith and Dr. F. Fincke, who brought the specimens to the laboratory. There was nothing of special note in the thoracic organs except that there was some adhesion of the layers of the pericardium.

The peritonæum was smooth. The stomach, intestines, and mesentery were removed together. The liver was not enlarged, was a little granular on section, and firm. The omentum was uninvolved. The stomach was free on its anterior wall, on the greater part of the posterior wall, and on the greater curvature, but was closely adherent at the pyloric zone to the contiguous parts. The organ was reduced in size, measuring in its extreme length thirteen centimetres; transverse diameter, from four to five centimetres. It was extremely firm and dense, and the tumor mass which was felt during life corresponded to it, and the hard, resistant edge corresponded to the greater curvature. The orifices were not narrowed; the walls were extremely thickened, from eight to ten millimetres at the cardia, and from thirteen to fourteen at the pylorus. The thickening was due partly to the great hypertrophy of the muscularis, but chiefly to the submucosa, which measured from three to five millimetres. The mucous membrane was uniformly smooth, excessively thin, and showed no erosions or ulcerations.

On microscopical examination, the mucous membrane was almost entirely deficient. The submucosa was occupied by large groups of cancer cells between strands of connective tissue. The layers of the muscular coat were much hypertrophied, and, invading the interstitial tissue and sometimes in the muscular bundles themselves, were numerous cancerous cells.

The pancreas was firm and hard and uniformly surrounded by thickened peritoneal tissue, infiltrated in places with cancerous new growth. The substance of the gland itself was normal. The mesentery was enormously thickened, measuring close to the root three centimetres. The peritonæum was thickened, presenting in places flat areas of carcinoma, and the mesenteric glands were uniformly enlarged and cancerous.

While we regarded this case as one of cancer of the stomach, we certainly were not aware of its remarkable character. The tumor so readily felt was thought to be an infiltration of part of the stomach wall, whereas in reality it corresponded definitely to the organ itself. The diffuse infiltrating carcinoma has to be carefully distinguished from the true cirrhosis of the stomach, consisting of fibrous overgrowth. The distinction can, however, only be made by the microscope, unless, of course, there is secondary infection of glands and neighboring organs. In this instance the lymph glands were infected, and I would call your attention particularly to the fact that the supraclavicular glands on the left side were also involved—a situation in which secondary tumors are sometimes seen in cancer of the stomach and of the œsophagus, and in which their presence may be of the greatest value in diagnosis. Not only may there be a primary infiltrating carcinoma of the stomach not distinguishable macroscopically from cirrhosis of the organ, but Dr. Welch, in showing this specimen at the Hospital Medical Society, called attention to an instance which he had reported of secondary infiltrating carcinoma of the stomach in a woman, aged forty years, who had double carcinoma of the ovaries. This beautiful plate, in Carswell's *Morbid Anatomy*, illustrates the condition very well.

LECTURE II.

NODULAR AND MASSIVE TUMORS OF THE STOMACH.

WE considered in the first lecture the cases in which the tumor was formed by the stomach itself, either in a state of extreme dilatation or extreme contraction. In twenty-one cases of the series nodular growths or diffuse thickening and infiltration were present; in three instances a massive infiltration. And first let me remind you of one or two anatomical facts. The only fixed portion of the stomach is the cardiac orifice, which is covered deeply by the left lobe of the liver, and externally corresponds to the seventh left costal cartilage near the sternum. The organ itself varies much in position with the degree of fullness or emptiness. The pylorus may be in the middle line, but when the organ is distended it is from six to eight centimetres to the right. It is usually, not always, covered by the liver. Fully two thirds of the stomach lie beneath the ribs in the left hypochondrium, and in contact with the abdominal walls are only part of the body and the pyloric region. Practically, however, we find that the organ is often depressed and so enlarged that a much more extended area than usually stated is exposed for palpation. Tumors limited to the cardiac orifice can not be felt at all, even when extensive. Those of the fundus and the posterior wall, and a considerable part of the lesser curvature, can only be felt when of large size. Tumors of a considerable extent of the greater curvature and a large section of the anterior wall are in accessible situations. It is of interest here to note the situation of new growths in the stomach, as determined in thirteen hundred cases analyzed by Professor Welch. The distribution was as follows: Pyloric region, 791; lesser curvature, 148; cardia, 104; posterior wall, 68; the whole or greater part of the stomach, 61; multiple tumors, 45; greater curvature, 34; anterior wall, 30; fundus, 19—so that at least three fifths of all tumors occupy the pyloric region.

The cases which have been under observation may be grouped into tumors at the pyloric region, tumors of the body of the stomach, and massive tumors occupying a very large area of the organ.

(a) TUMORS OF THE PYLORIC REGION.—Of the twenty-four cases, there were seventeen with a tumor mass of some size or form to be felt at the distal portion of the organ. In ten of these dilatation of the stomach was present, prominent enough to itself cause a tumor and have been considered in the first lecture.

Before entering upon a description of the cases an important question arises: Is the normal pylorus palpable? It may be answered, I think, in the affirmative, with certain qualifying conditions. The pylorus forms a definite ring-like muscular valve, readily to be seen and felt in the exposed organ. Whether, as has been stated, it relaxes and contracts rhythmically at definite intervals has not been fully determined, but I would remind you of the statement made by Beaumont, in his experiments on the movements of the stomach of St. Martin, that when the thermometer was placed toward the pyloric orifice it was at first firmly grasped, and then, by gentle relaxation, allowed to pass. If the stomach be exposed in a cadaver and a couple of towels laid upon it, on palpation over them the pyloric ring is readily felt. So also, I believe, it may sometimes be detected during life. Though normally covered by the anterior margin of the liver, it is freely exposed in a very considerable number of cases, and when the stomach is depressed or in a state of atony the pyloric ring is always below the edge of the liver. In persons with very thin walls, particularly in cases of enteroptosis in women, palpation in the boundary of the epigastric and umbilical regions may discover a small, transversely placed body, varying in position, with respiration which sometimes gives the impression of a structure alternately in contraction and relaxation. In some cases it may even be rolled beneath the finger. At intervals gas is felt to bubble through it. From the pancreas, which is also sometimes palpable, it is readily distinguished by the alternate relaxation and contraction, and by the bubbling of gas through it. The condition is one of some importance, as it may lead to the suspicion of gastric cancer. Thus I saw with Dr. Salzer, in September, 1892, a woman aged thirty-two, a chronic dyspeptic, but who lately had had very severe symptoms, and had lost rapidly in weight, having fallen from ninety-five to sixty pounds. At the junction of the epigastric and umbilical regions, a little to the left of the middle line, there was a soft, cylindrical structure, which descended with inspiration. Its transverse extent was not more than three or four centimetres. It hardened definitely under palpation. Gas was felt escaping through it. The patient was, of course, extremely emaciated, and the discovery of the tumor, together with the pronounced stomach symptoms, led to a suspicion of malignant disease. The case was subsequently under my care for seven months, and proved to be an extremely obstinate form of anorexia nervosa; but she gained in weight from sixty to a hundred and fifteen pounds, and the improvement continues to date.

I believe that in very thin-walled persons, particularly those with atony of the stomach, the pylorus, i. e., the ring and adjacent part, may sometimes be felt as a narrow, tubular structure, the distinguishing features of which are the alternate relaxation and contraction and the bubbling of gas through it.

Of the seventeen cases, two were instances of cicatricial thickening and stenosis, and in fifteen there was either a cylindrical tumor or a nodular mass.

A tumor-like formation at the pylorus may be due to a number of causes—cicatricial contraction and thickening about an ulcer, hypertrophy of the pylorus, and cancerous growths—all of which conditions may lead to stenosis of the orifice and secondary dilatation of the stomach. Again, the first part of the duodenum and the pylorus may be invaded by growths from contiguous organs, as in a case to be subsequently mentioned, in which the tumor in the pyloric region was caused by invasion of the duodenum by a cancer of the colon. And, lastly, there may be mentioned as a cause of dilatation of the stomach, stenosis of the pylorus by dislocation. Thus adhesions may form between the gall bladder and the pylorus, and this portion of the organ is drawn up and the orifice narrowed. A remarkable instance of the kind was operated upon in the hospital by Dr. Finney in August last.

Tumors of the pylorus are usually, but not always, associated with dilatation; thus there were only four out of the seventeen cases in which the organ was not distended. The cases of pyloric growths or thickening may be grouped as follows:

Thickening and Induration from Healing of an Ulcer.— Two cases come in this category—Case X of the series already given in the first lecture, which presented a very greatly dilated stomach. There was a ridge-like mass, freely movable, to be felt midway between the umbilicus and the right costal margin. The prolonged history of dyspepsia, the moderate wasting, the fact that she had on several occasions vomited blood, and the small size of the pyloric tumor, suggested cicatrization about an ulcer. In the following case gastro-enterostomy was performed by Dr. Finney, and, unfortunately, on the tenth day the patient died of an acute colitis. The nodular tumor was very well defined, particularly after the stomach had been emptied.

CASE XII. *Dilated Stomach; Tumor Mass at the Pylorus; Gastro-enterostomy; Autopsy; Stenosis from Ulcer.*—Mary G. aged twenty-two, colored, admitted on July 29th, complaining of a "gnawing in the stomach" and "vomiting spells." Family history good. Was healthy as a young girl, with the exception of a slight attack of pneumonia about eight years ago, and malaria a year later.

Her present illness began last April with loss of appetite and weakness, and she began to lose flesh. About two months ago she noticed a lump in the abdomen, which she thinks has got larger. About this time she began to vomit at irregular intervals, without any nausea or acid eructations. The attacks gave her relief, and she says the lump seemed smaller after them. The vomitus was copious, greenish in color, and watery. Bowels have been constipated. The feet have swollen sometimes. She has never vomited blood.

Condition on Admission.—Medium-sized, greatly emaciated, lips and mucous membranes pale, tongue presents a whitish fur. Pulse 92, regular, of fair volume. With the exception of a soft systolic murmur at the heart apex, there are no abnormal physical signs in the thoracic organs. The abdomen is symmetrical, not specially distended. In the right hypochondrium, just below the costal margin and opposite to the cartilage of the eighth rib, there is a small nodule, apparently the size of a horse-chestnut, which descends with inspiration and gives the impression of being at the pyloric orifice. It varies somewhat in position and in firmness; thus the day on which I examined her (September 1st) it could by no means be satisfactorily determined, although the day before Dr. Thayer had been able to feel it with the greatest distinctness. On inflating the stomach, the area of gastric tympany was found to be greatly increased, extending from the fifth rib above fully three fingers' breadth below the umbilicus. During great distention of the stomach the nodular mass could not be felt. The material obtained after a test breakfast was mixed with fragments of undigested bread and curds of milk which had been taken the day before. It has a strong sour smell, suggesting butyric acid. The reaction was acid. The congo and tropæolin tests were negative. Uffelman's test gave sharply positive results.

After treatment for some time with washing out the stomach no special benefit followed, and, as the lesion seemed most probably stenosis from ulcer, it was thought advisable to attempt dilatation. Accordingly, Dr. Finney made an exploratory examination and found externally much thickening about the pylorus. He opened the stomach on the anterior wall, and, exploring digitally, found the orifice much narrowed, partly by contraction, partly by polypoid excrescences, several of which were removed. As it was doubtful whether this would be sufficient, a communication was made between the stomach and the jejunum. The patient did very well for ten days, but then vomiting and diarrhœa set in and the latter became severe and she died in the third week after the operation.

The post-mortem showed chronic adhesive peritonitis about the pylorus and over the surface of the liver. The jejunum was very firmly adherent to the anterior wall of the stomach. The tumor mass which had been felt during life was the thickened pylorus. When laid open, a large ulcer was found in the pyloric region with much puckering of the mucosa about it and cicatricial contraction.

Tubular and Small Nodular Tumors at the Pylorus.— Next let me call your attention to four cases which have presented a good deal of difficulty in diagnosis, not as to the existence of the tumors, but as to their nature. In three there was a cylindrical, somewhat tubular-shaped tumor to be felt. In two of them there was evidence of some dilatation of the organ after inflation, but the symptoms in each case were those of chronic dyspepsia, not of extreme dilatation of the stomach. I will first read you the report of the cases.

CASE XIII. *Chronic Dyspepsia; Cylindrical Tumor of the Pylorus.*—Mr. S., aged seventy years, was admitted to Ward C October 6th, complaining of dyspepsia.

The patient has been a dyspeptic ever since 1843, and unless very careful with his diet had fullness and tenderness in the epigastrium. He frequently had very disagreeable feelings after eating, and would vomit or regurgitate the food. In spite of the dyspepsia, he has always been robust and has never been laid up in bed. For the past two years the dyspepsia has been more troublesome and he has frequently had pain after eating. The gastric trouble increased so much that four months ago he had to take peptonized milk. There has been no vomiting, though at any time he could regurgitate the food. For the past ten months he has lost in weight (as much as twenty pounds) and in strength, and has been in very low spirits.

The patient is a well-preserved man for his years, of spare habit, but neither emaciated nor cachectic. The general physical examination is negative. The heart, arteries, and lungs are normal.

Abdomen flat, on palpation soft, no pain. Four centimetres above the navel there is to be felt a ridge-like tumor, which can be rolled beneath the fingers, and which extends six centimetres in the transverse direction. On inspiration it descends slightly. It can be moved up and down; the surface is smooth, and firm pressure is not painful. The point of greatest interest is the remarkable variability in consistence within a few minutes. At times it is firm, hard, and ridge-like, and within a minute it becomes very much softer. Gas can be felt to bubble through it.

There is no glandular enlargement. After a test breakfast sixty cubic centimetres of a light greenish fluid, rather slimy and mucoid, were removed, which gave none of the reactions for free hydrochloric acid.

Patient left hospital on October 22d, somewhat better, but I hear from Dr. H. M. Thomas that he died before Christmas.

CASE XIV. *Dyspepsia of a Year's Duration; Cylindrical Tumor of Pylorus.*—Bertha N., aged forty-four, admitted to Ward G, October 28th, complaining of pain in the epigastric region and loss of appetite.

Father died of tuberculosis; mother of hæmorrhage from the uterus. She has had two healthy children.

Her present illness is of more than a year's duration. She has had weakness, loss of flesh, and for several months the appetite has been very poor. She gives an account of what she calls spasms, which would appear to be fainting fits at the menstrual period. After eating she has pain and has to lie down, and finds relief by pressure on the abdomen. There has been no vomiting. Lately she has lived principally on milk.

Present Condition.—There is uniform pigmentation of the face except in one or two spots on the cheeks and chin. The general pigment over the surface of the body seems to be somewhat increased, particularly on the abdomen and the arms. She is decidedly emaciated, but the lips are of a red color, and she has not a cachectic look. The tongue is furred. The examination of the thoracic organs is negative. The heart sounds are clear. The superficial arteries are slightly thickened.

The abdomen is flat, somewhat sunken, walls relaxed. In the epigastrium, a little below the ensiform cartilage and extending into the right hypochondriac region, a cylindrical tumor can be felt and rolled beneath the fingers; at intervals gas can be felt to bubble through it. The inflation of the stomach shows the organ to be depressed and somewhat dilated.

An hour after a test breakfast there was very little material obtained. It was acid to litmus paper ; no reaction to congo paper. On washing out the stomach, two or three lumps of bread, not digested, were obtained. There was no free hydrochloric acid. Subsequent examinations showed persistence of this ridge-like mass in the epigastrium, but it varied considerably in position; thus on November 4th it could be felt distinctly to the left of the middle line, and even a ridge beneath the skin could be seen in this position to descend on deep inspiration. The patient left the hospital December 22d, and has not since reported for examination.

CASE XV. *Dyspepsia for Several Years; Cylindrical Tumor of the Pylorus.*—M. O., aged fifty-four years, admitted March 10, 1893, complaining of pain in the stomach. Father died at sixty-two of pulmonary tuberculosis, of which disease also one brother died. Mother died at sixty-five of dropsy.

She has always been healthy; married; has three children. Has been subject to dyspepsia for several years.

Last winter she had a great deal of pain in the epigastrium, diarrhœa, and vomiting of a greenish fluid. She was in hospital for eleven weeks, but improved, and from July of last year until January she was perfectly well and could eat anything, but she has been paler and more sallow and has lost in weight.

The present trouble began about four weeks ago with pain in the epigastrium and great tenderness. She has had nausea, but no vomiting. The patient is a small woman, looks ill, complexion very sallow, lips and mucous membranes pale. Tongue is flabby but clean. With the exception of a soft murmur at the apex of the heart, examination of the thoracic organs is negative.

The abdomen is a little full, everywhere tympanitic. There is a marked depression just below the costal margin on both sides; no peristalsis visible. Palpation at first caused so much tenderness that the examination was unsatisfactory; but as she drew a deep breath a small nodular mass could be felt descending beneath the ribs in the parasternal line. In a few days she was somewhat better, and a more thorough examination could be made. In the parasternal line, midway between the costal margin and the navel, there is a cylindrical mass, transversely placed, which can be rolled beneath the fingers. It is extremely sensitive; no flatus is felt passing through it. On deep inspiration, it descends and the fingers can be placed above it so as to hold it down. On percussion, it is resonant. After inflation of the stomach there is no marked dilatation of the organ; no peristalsis is seen, but the tumor is then not so easily palpable.

Ewald's test breakfast, withdrawn fifty minutes afterward, showed about fifty cubic centimetres of grayish, slimy fluid, containing portions of undigested food, and, on testing, no free hydrochloric acid.

Patient was under observation until May 11th. The tumor did not change in any way. The pain lessened and she gained in weight from a hundred and seventeen pounds on admission to a hundred and twenty-five pounds on discharge.

CASE XVI. *Nodular Tumor in the Pyloric Region: Dilatation of Stomach.*—J. A. R., Talbot County, Md., seen October 19, 1892, with Dr. Chamberlaine. The patient had been admitted to Ward C, November 17, 1891, with the following history:

For ten months he had had occasional paroxysms of boring pain in the abdomen, coming on usually at night. Six months previous to admission he had noted a small nodular tumor in the abdomen, which he says has gradually become larger. He has had no symptoms of indigestion or special distention of the stomach after eating; has vomited only twice, on both occasions, he thinks, caused by a severe paroxysm of pain. He has lost about fourteen pounds in the past month. He was a man of spare habit, but was not anæmic. A test breakfast, withdrawn an hour after the meal, showed about two ounces of fluid containing small lumps of partly digested food. Free hydrochloric acid was present.

In the abdomen, just above and to the right of the umbilicus, there was felt a rounded tumor about the size of an English walnut, freely movable. On inflation, the stomach tympany extends two fingers breadth below the umbilicus.

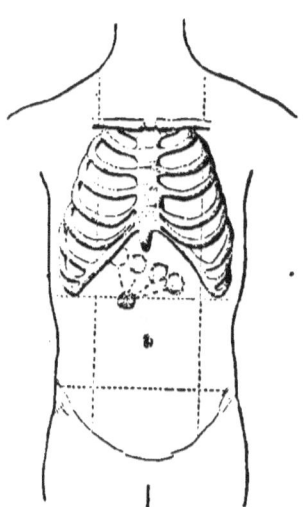

FIG. 18.—Positions into which the tumor could be moved in Case XVI.

The patient remained in hospital for a couple of weeks; had no special gastric symptoms, gained in weight, and returned to his home November 27th. The case was regarded as one of tumor of the pylorus, and he was told if the trouble increased an operation might be advisable.

October 19, 1892.—Patient examined to-day; has been very much better; entirely free from pain; has had no vomiting; has been taking an ordinary diet; no nausea; no sense of distress after eating. He looks and feels well.

The abdomen is a little full in the umbilical region, flat in the epigastric. It is everywhere soft and painless; nothing can be felt in the epigastric region. Midway between the navel and right costal margin there is the same well-defined, firm, hard nodule to be felt, which is now painless. It descends with inspiration and can be moved about in the positions noted in the diagrams. The patient says that it is not nearly so evident when the stomach is empty. It is prominent enough to be seen when the skin is pressed over it. It can be pushed far up under the right costal margin and at first could not be felt, as it was high in this position and only made to descend by deep inspiration. To the left it can be pushed beyond the middle line to a point midway between the navel and the left costal margin. It is very mobile. On inflation of the stomach the lower border was found to descend some distance below the umbilicus—three or four fingers' breadth—while the lesser curvature was almost as low as the umbilicus. When the stomach was inflated the mass could not at first be felt, but afterward was found a trifle more to the right and not apparently quite so superficial as it was before the distention. Nearly a year has elapsed since this patient left the hospital; in that time the nodular tumor has increased but little in size, and the patient's general condition is remarkably good.

In Cases XIII, XIV, and XV the tumor had a definitely cylindrical shape, and in Case XIII there seemed to be no question that the tumor was a thickened pylorus, as marked variations occurred in its consistence and gas could be felt bubbling through it. So also in Case XIV similar features seemed to indicate clearly that the tubular structure, so readily felt, represented in reality the thickened pyloric ring and adjacent part of the stomach. In Case XVI the tumor was rounded, nodular, and very movable. It did not vary in consistence and no gas was felt to bubble through it. It felt very hard and firm. While its local features seemed to indicate definitely that it was a new growth, the general condition of the patient after its existence for eighteen months seemed to be very much against the view that it was cancer. Scirrhus, however, may develop very slowly indeed at the pylorus, and make very slight progress within six months. Thus in Case XVII, on the first admission in September, after an illness of fifteen months' duration, the pyloric tumor constituted a tubular, sausage-like tumor, which could be rolled beneath the finger. I frequently discussed with Dr. Thayer the question whether it was an instance of hypertrophic thickening of the pylorus or a scirrhus growth, and the time element seemed to be in favor of the former; but when the patient returned in April of this year the tumor, which meantime had increased in size, was found to be a scirrhus. The hypertrophic stenosis, with which the annular scirrhus of the pylorus could alone be confounded, occurs in connection with chronic gastritis, with scars of old ulcers, in connection with a cirrhosis in other parts of the stomach or intestines, and sometimes with a general sclerosis of the tissues of the mesentery and peritonæum. It may be impossible, for a time, to give a positive opinion. In either case, however, the condition is serious.

Cases of Large Nodular Growths at the Pylorus.—These constitute a large majority of pyloric tumors. Most of the cases have already been described in the first lecture in connection with the dilatation of the stomach. In some the tumor itself was visible beneath the skin. The following cases are good illustrations of this type of growth. In one the tumor was an annular cancer, which was removed by operation:

CASE XVII. *Annular Carcinoma of the Pylorus; Excision of Growth; Death; Autopsy.*—Henry M., aged sixty-one years, laborer, admitted September 30th, complaining of "sour stomach" and vomiting.

No history of hereditary disease. Patient has been healthy and strong, and has had only a few illnesses. He has used alcohol freely, but has not been a heavy drinker.

Present illness began fifteen months ago with an uneasy feeling in the abdomen and churning sensations, which continued until he vomited a watery, very bitter fluid, which sometimes had a greenish-yellow color. At first his appetite and digestion re-

NODULAR AND MASSIVE TUMORS OF THE STOMACH. 47

mained good. The uncomfortable feeling after eating gradually increased, and during the past five months, although his appetite has been good and he took his regular meals, vomiting came on about an hour and a half afterward. He has never brought up very large quantities, and pain has never been a prominent feature. He has lost weight rapidly, and within six months has fallen from a hundred and fifty to a hundred and fourteen pounds. The bowels have been very irregular, and he sometimes has had no movements for a week or ten days, and recently has gone as long as sixteen days.

Present Condition.—Large-framed man; much emaciated, particularly in the face. The cheek bones are prominent and the eyes sunken. The mucous membranes are not specially anæmic, and the facies can scarcely be termed cachectic. The emaciation is marked about the thorax; the skin is smooth and clean; superficial lymph glands are not involved. The tongue is a little swollen, indented, and furred.

The abdomen is flat, very much below the level of the costal margin. There is a slight prominence just to the right of the navel, and on deep inspiration a ridge-like mass descends below the point.

On palpation, there can be felt just above and extending to the right of the navel a firm mass which descends on inspiration and can be rolled beneath the fingers, giving one the impression of a tubular, sausage-like tumor. On deep inspiration, it moves down nearly three inches and can be readily held at the navel, and then slips away from beneath the fingers.

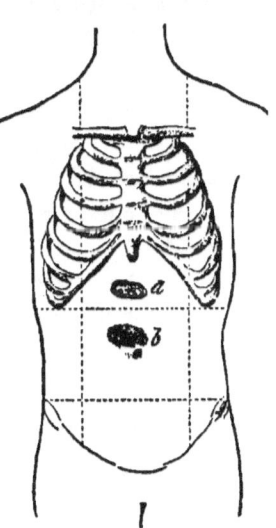

FIG. 14.—*a*, position of the tumor in expiration; *b*, in inspiration.

The patient was placed upon a careful diet of milk and egg albumin, upon which the nausea disappeared and he became very much more comfortable. The examination on the 17th showed that the elongated mass above referred to had changed in position and lay to the left of and just above the navel.

The liver is not enlarged ; the edge of the spleen is not palpable.

The patient got dissatisfied with the diet, and left the hospital October 17th, though he seemed to be considerably improved.

April 6, 1893.—Patient returned to the hospital, having had, in the space of nearly six months which has elapsed since the last note, very marked gastric symptoms and occasional attacks of vomiting of large quantities of fluid. He has lost in weight, though he does not look much more emaciated than when he left.

Examination.—Abdomen is scaphoid and the walls held rather rigidly. Between the ensiform cartilage and the navel the solid rounded tumor present in September can be felt in the same position, but it appears definitely to have increased in size. It can be rolled beneath the fingers, and part of it at least varies somewhat in consistence, becoming harder and firmer. The stomach is moderately dilated, and when inflated with gas shows distinct peristalsis, and then the tumor is not nearly so evident.

The advisability of an operation was suggested to the patient in the autumn, but he refused ; now he is anxious that one should be performed.

The age, the profound emaciation, and the very evident increase in the size of the tumor suggested cancer rather than cicatricial contraction, with thickening at the pylorus ; and as the case was a desperate one, and the man's condition hopeless, the election of operation was left to the patient.

On the 11th Dr. Halsted operated and found a solid annular growth in the pylorus, extending for about seven centimetres, reaching to the orifice, but not extending into the duodenum. As there were no special adhesions and no nodules, he proceeded to resect the growth, which was done successfully. The operation lasted about two hours, and the patient was very much exhausted ; he rallied well through the night, and seemed very comfortable, but failed rapidly on the 12th and died on the 13th.

CASE XVIII. *Cancer of the Stomach ; Large Tumor in the Pyloric Region.*—Patrick K., aged twenty-eight years, admitted

NODULAR AND MASSIVE TUMORS OF THE STOMACH. 49

to Ward E, January 3, 1893, complaining of pain in the abdomen. Patient had been an orderly in the hospital in 1889–'91, during which time he was laid up on several occasions—once in December, 1889, with vomiting and diarrhœa ; again in April, 1890, with acute gastritis, an attack associated with much pain ; and in April, 1891, he had a sharp attack of amygdalitis. A very noticeable feature was the persistent anæmia, and on several occasions in 1889 and 1890 I had examined his blood, which presented all the characters of a secondary anæmia of moderate grade. From his earliest boyhood he has been subject to nose-bleeding, and has always, he says, been pale.

In October, 1891, he returned to his home in Ireland, and remained fairly well, but was troubled on several occasions with epistaxis. He returned to this country last year. Four months ago he had an attack of pain in the abdomen with vomiting, and these symptoms have persisted ever since. The vomiting is chiefly after taking food, and the pain is also most severe at this time. He has never vomited blood. Bowels have been constipated. He has lost in weight, he thinks, as much as fourteen pounds. He is short of breath on exertion, and when he walks about for any length of time the feet and ankles swell.

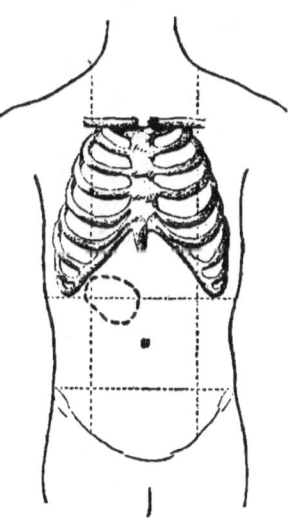

FIG. 15.—Situation of the tumor in Case XVIII.

Present Condition.—He is very anæmic, but not emaciated; his face is full; blood count, 3,000,000; hæmoglobin, thirty per cent.; eyelids a little puffy; hands very pale; pulse, 87, soft, compressible; radials a little thickened; vessels of neck throb; the heart sounds are loud and clear at the apex, the second very ringing and accentuated at the base; no murmur. The examination fo the lungs was negative.

Abdomen full and a little prominent; on palpation, everywhere soft and painless until the right epigastric region is reached. Here, under the costal border in the parasternal line, there is a resistant mass which extends to the right almost as far as the middle line and to the left as far as the nipple line, and below at least six centimetres from the costal border. During a deep inspiration the mass descends and the fingers can then be placed between it and the costal margin. In the middle line in the epigastric region nothing is palpable. There is resonance over the above-described tumor mass. There is no peristalsis apparent; no gurgling to be felt in the mass. After dilatation of the stomach the tympany in the parasternal line was at the seventh rib, and extended two fingers' breadth below the navel. The tumor was pushed far over nearly beyond the nipple line.

A test breakfast withdrawn an hour after gave a hundred and twenty cubic centimetres of thick, dark-brown fluid, containing undigested food and a few shreds of clotted blood. The reaction was acid; there was no reaction with congo paper, nor with the other tests for free hydrochloric acid. Uffelman's test for lactic acid was positive; starch test negative. The spleen was not enlarged; the liver not enlarged. The urine presented no changes.

The patient failed rapidly, became very anæmic, and lost sixteen pounds in weight within a month. The vomiting was very troublesome and intractable. No special change took place in the character of the tumor mass, though as he became thinner it was rather more evident. He died on February 23, 1893. There was no autopsy.

Practically, then, the tumors at the pyloric orifice which we have been studying consisted of cicatricial thickening caused by ulcer, possibly hypertrophic stenosis, annular carcinoma, and large nodular masses. There are one or two points of general interest to which I will here refer. In the first place, the tumor is always larger than you expect from the examination through the abdominal wall. This has to be borne in mind in a discussion on the advisability of operation. It is frequently very variable—well and plainly to be felt to-day, and perhaps scarcely palpable to-morrow—variations which depend a great deal upon the degree of dilatation of the organ, particularly of the portion known as the pyloric pouch, which may cover over and mask even a large pyloric tumor. Examination in the knee-elbow position often gives valuable information as to the relations and positions of a tumor, and should never be omitted in doubtful cases. The value of careful palpation with a view of determining whether gas bubbles through the tumor is of the very greatest importance. The masses are usually firm, hard, and often of a stony consistence; sometimes the nodular masses formed by the glands in the neighborhood of the pylorus can be very plainly felt. A feature in the pyloric tumor which merits special attention is the mobility. I have already referred to it in Case V, in which the solid rounded tumor mass could be pushed beneath the ribs on the right side, far down into the iliac regions, and far over to the left costal border. So also the nodular tumor in Case XVI, which I have just read to you, was extremely movable. I reported a case a few years ago (*University Medical Magazine*, vol. i, p. 368) which is of great interest in this connection:

The patient, aged sixty-five years, was admitted to the Philadelphia Hospital, October 14, 1888, with chills and fever. The blood, however, was negative, and it was ascertained that he had had for some weeks distress after eating, and our attention was then directed to a more careful examination of the abdomen. On November 11th the following note was made : Patient is anæmic and emaciated; the abdomen flattened; there is a prominent projection below the left costal border in the parasternal line reaching nearly to the navel and descending with inspiration. Palpation reveals a firm, hard mass, occupying the left hypochondriac region and the left half of the epigastric region. It is smooth and not painful, and can be moved from side to side to an extent of two or three inches. Percussion over it gives a flat tympanitic note; liver dullness not increased; glands in the groin are double the normal size; vomited matters are brownish in color, acid, but contain no sarcinæ. The tumor changed curiously in position from day to day; at one time it was far over in the right hypochondriac region, entirely beyond the middle line, but more commonly a greater portion of its extent was in the left hypochondriac region. On several occasions it seemed to have disappeared altogether, and only a hard, small mass could be felt far over in the left hypochondrium. Patient sank gradually and died November 20th. When the abdomen was opened no trace of tumor was visible until the stomach was pulled down and to the right. It was then seen that the mass had fallen back into the left hypochondriac region below the ribs, where it was completely covered by the splenic flexure of the colon. The duodenum was curiously elongated and straightened; from the pyloric ring it measured over two inches as a straight tube. The pancreas was also drawn over to the left. The tumor could readily be pushed to occupy the positions in which it was felt during life. It involved the anterior wall of the stomach, which, when opened, presented a large hemispherical mass, involving three fourths of the circumference of the pyloric region and extending to within an inch and a half of the ring. The surface of the mass was ulcerated, and at the base near the greater curvature suppuration had taken place.

(*b*) TUMORS OF THE BODY OF THE STOMACH.—Tumors of the pyloric region often encroach extensively on the anterior wall of the stomach, but I have placed in this category three cases in which the tumor mass appeared to be more in the central part of the organ. In Case XIX the left epigastric region was occupied by a rounded, irregular tumor, and the patient had had marked gastric symptoms and had vomited blood. Though there was no question as to the nature of the growth, it is interesting to note that during his stay in hospital he gained six pounds in weight. In Case XX the tumor mass was more extensive and seemed to involve a large section of the anterior wall of the stomach, forming a very prominent and readily palpable tumor. In Case XXI a large nodular mass could be felt between the left costal margin and the navel. It was unusually firm, and post mortem showed that it occupied more than ten centimetres of the anterior wall of the stomach.

CASE XIX. *Large Tumor of the Body of the Stomach.*—Gustave P., a shoemaker, aged fifty-three years, admitted December 28, 1892, complaining of pain in the abdomen and back.

Parents died over eighty years of age ; one sister died of cancer of the womb.

Patient was born in Germany ; has been very healthy ; has not been a heavy drinker ; and denies lues.

The present illness he dates as far back as eight years ago, at which time he had dyspeptic symptoms, which persisted for two or three years ; then he remained quite free from them for about three years, but early in 1890 they recurred. He has had uneasy sensations after meals, and belching, sometimes bringing up acid fluid. During the past summer he had a good deal of vomiting, and once in June brought up dark-brown fluid, which was said to be blood, and the next day the same material was noticed in the stools. His appetite is fair, but he is afraid to eat, and lately has only been taking liquids. He has not lost very much in weight — only about five pounds in the last six months.

Present Condition. — Patient is emaciated, pale, and a little sallow ; mucous membranes distinctly anæmic. Tongue has a patchy coating and indented edges.

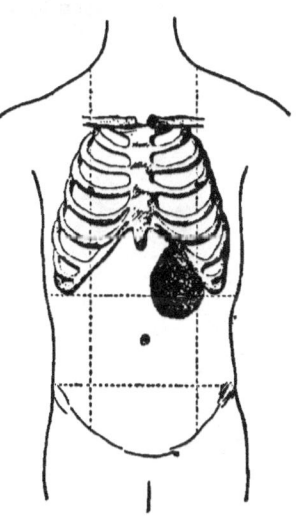

FIG. 16.—Outline of the tumor mass in Case XIX.

Examination of the thoracic organs is negative.

The abdomen is symmetrical, a little depressed below the costal border ; no peristalsis visible. On palpation, the left epigastric region is occupied by a superficial mass with a rounded, irregular, nodular surface. It extends sometimes almost to the middle line, and below crosses the transverse costal line. To the left it extends to the nipple line. It is a little painful on firm pressure ; descends with inspiration. The pulsation of the aorta is transmitted through it. Percussion gives a flat tympany over the mass. Ewald's test breakfast, withdrawn fifty-five minutes after, yielded about seven hundred cubic centimetres of brownish fluid with a heavy sediment of undigested food ; odor acid. The filtrate turned blue litmus red, congo red to blue, and yielded a rose-red color with phloroglucin vanillin. Uffelman's test negative. The urine was normal. The patient remained up and about the ward, and with a careful diet was made much more comfortable. The condition of the gastric juice was frequently tested ; thus on January 4th Ewald's test breakfast, withdrawn an hour later, yielded about five hundred cubic centimetres of sour, yellowish food matter, which gave the reactions previously noted.

The patient continued to improve, gained in weight from a hundred and fourteen to a hundred and twenty pounds, and was in every way more comfortable. He had almost constantly, while in hospital, a little fever, temperature rising to 100°, sometimes to 101°, every day. No special change occurred in the position or condition of the tumor mass. He was discharged February 27th, and has not since been heard from.

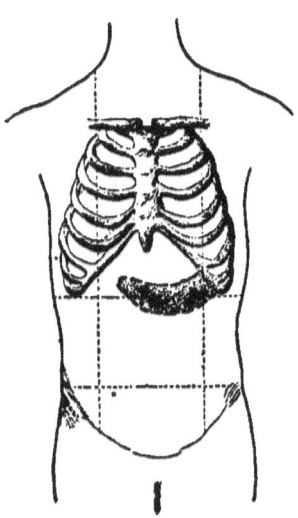

Fig. 17.—Area of the tumor mass in Case XX.

CASE XX. *Tumor of the Body of the Stomach.* — N. R., aged sixty-nine, shoemaker, German, came to this country in 1872. Admitted March 13th, complaining of loss of appetite, nausea, and vomiting.

Patient was sickly as a child. Six years ago he had a fever which kept him in bed for three weeks. He had been a moderate drinker ; denies syphilis. Has not been a dyspeptic.

Present illness began before Christmas with pain after eating, nausea, and vomiting, the latter usually a short time after taking food. The appetite has failed and he has lost rapidly in weight. He has had much fullness and distress in the epigastric region, but no very sharp pain. Of late all these symptoms have become aggravated. He has never had vomiting of large quantities of food.

Present Condition.—Patient is a large man, still fairly well nourished. The lips and mucous membranes are pale; tongue has a thick white coat. Pulse is regular, 64, tension slightly increased the vessel wall sclerosed. Temperature is normal. Examination of the thoracic organs is negative.

The abdomen is symmetrical, except that there is a slight prominence at the end of the tenth rib on the left side; on palpation, soft, and nothing is felt until the epigastric region is reached. In the region indicated in the figure is a mass which moves freely in inspiration. The lower border is sharp like that of the liver or spleen; the surface is irregular and somewhat nodular. On percussion, there is a distinct tympany over the mass. On inflation, there is no abnormal dilatation of the stomach.

Ewald's test breakfast given at 8 A. M.; at 9.30 a tube was inserted. There seemed to be some slight obstruction about the cardiac orifice, and about fifty cubic centimetres of coffee-colored fluid removed, together with a very little fresh blood. There was no free hydrochloric acid. Microscopically, it presented fresh blood-cells, blood-pigment, and remnants of food. No enlargement of the lymph glands.

No material change took place within two weeks in the patient's condition; he was evidently failing, and he decided to go home.

CASE XXI. *Tumor of the Pyloric Region and Anterior Wall; Perforation; Peritonitis.*—August B., aged fifty-eight years, farm laborer, German, admitted complaining of pain in the abdomen, loss of appetite, vomiting, and insomnia. Father died aged fifty-six, cause unknown; mother, of dropsy at sixty; one sister died of cancer of the stomach; no history of tuberculosis in the family.

Has always been a healthy man; the father of seven children. Has been a moderate drinker; denies venereal disease. Has always had good digestion; never suffered from dyspepsia.

His present illness began four weeks ago with pain in the

abdomen and vomiting, which comes on very shortly after eating. He has never vomited any great quantities. Though he does not appear to have had any marked stomach symptoms, during the past six months he has lost in weight from a hundred and sixty-five to a hundred and thirty-five pounds. Patient is a medium-sized man, pale, thin, lips and mucous membranes pale, and he looks somewhat cachectic. Thorax is symmetrical; above the left clavicle one of the lymph glands is enlarged and hard. Examination of the lungs and heart negative.

The abdomen is full, particularly in the right epigastric region. Here, on palpation, a nodular mass can be felt midway between the costal margin and the navel. It is flat and extends transversely as far as two cubic centimetres beyond the median line. It is hard, a little painful, and descends with each inspiration. It is resonant on percussion. No peristalsis was felt, no changes in consistence, and no gas was felt to pass. The stomach tympany begins at the seventh rib in the left parasternal line and does not quite reach the navel. After inflation of the organ no peristalsis is seen.

Test breakfast was given and a tube inserted an hour afterward. The fluid obtained was dark-brownish, with a sour odor; contained organic acids, but no free hydrochloric. The blood count showed hæmoglobin forty-five per cent., and the red blood-corpuscles about four million.

The patient left the hospital and was readmitted April 12th, and the following notes were made: He is very emaciated and looks cachectic. There is a marked prominence in the epigastric region, just below the ensiform cartilage, and here very slight irregular movements may be seen. The indurated mass noted above appears to have increased in size. It lies at the junction of the umbilical and epigastric regions, and, on inspiration, descends almost to the navel. It is firm and resistant. There is no peristalsis visible after inflation of the stomach, and no change in the position of the tumor. Just above the navel in the linea alba there is to be seen a flattened prominence, which feels soft and like a little fatty tumor beneath the skin. Patient became progressively weaker and died June 4th.

Autopsy.—Peritoneal cavity contains nine hundred cubic centimetres of turbid fluid; fibrinous exudate covers the intestines. There is a large tumor mass to the left of the pylorus, involving the anterior wall of the stomach nearly to the cardiac end. Midway between the greater and lesser curves is an oval perforation, measuring seven by three millimetres, through which the contents of the stomach can be squeezed. On opening the stomach, there is a large ulcerated cancer extending laterally for ten centimetres. The stomach walls in the neighborhood of the ulcer are much infiltrated and are raised and in places overhang the ulcer. In the anterior wall there is the perforation already mentioned. The ulcer does not extend to the pyloric ring. The glands about the stomach and pancreas are enlarged. The head of the pancreas is also involved. There are small white tumor nodules on the surface of the omentum and mesentery.

Death from perforative peritonitis is not a very uncommon complication of cancer. Perforation may also take place externally. A more common communication is with the colon, which in all probability took place in Case XXIV. A rare perforation in cancer, which I do not see mentioned even in the exhaustive article of Professor Welch, is into the pericardium, which I found at autopsy in a case of the late Palmer Howard's, of Montreal. There was, of course, the most intense pericarditis, and the group of physical signs of pneumopericardium.

(c) MASSIVE TUMORS OF THE STOMACH.—No cases are more difficult to recognize than those in which the walls of the stomach are extensively infiltrated. You might think that under these circumstances the diagnosis would be made with the greatest ease, but in reality they are cases which require no little care and study. Of the three cases which I shall narrate to you, in one the diagnosis was easy and definite; in one the tumor was so extensive, occupying such a large area in the left side of the abdomen, that some doubt existed as to whether it was not associated with the spleen or the kidney, while in the third there remains a doubt as to the exact nature of the growth.

CASE XXII. *Cancer of the Stomach; Prominent Tumor in the Epigastric Region.*—H. P. C.. aged fifty-seven years, admitted September 1, 1892, complaining of loss of appetite, progressive weakness, and irregular pains in the abdomen. Family history is good; parents lived to old age.

Patient since childhood has been healthy; appetite always good until about six years ago, when he began to have dyspepsia, occasional attacks of nausea and eructations, and sometimes vomiting. He appears at this time to have had considerable gastro-intestinal disturbance, as he had also diarrhœa, and became very weak and emaciated—quite as much so, he says, as he is at present—and the ankles were also swollen. The vomiting and nausea and diarrhœa stopped and he gained in weight, but his digestion for the past five years has not been as strong as before the illness six years ago. In June of this year he began to have feelings of oppression in the stomach, as though he had eaten a very full meal, but he had no vomiting; once or twice the food has been regurgitated. A marked symptom at first was inability to swallow after the second or third mouthful. With these local symptoms there has been loss of appetite and progressive emaciation. He has lost forty pounds since June. He has never had any severe pain; only a feeling of heaviness and distention after eating.

Condition on Admission.—Patient is a large, well built-man; face pale, and he looks depressed. There is marked emaciation. The pulse is of fair volume and the heart's action is strong. The temperature during the sixteen days he was under observation was always a little elevated toward evening, rising on several occasions to 102°. He had no cough; respirations quiet.

Abdomen flat, except in the epigastric region, in which it is a little prominent. On palpation, it is soft, painless, and just below the ensiform cartilage, extending across the whole upper zone, is a firm, resistant mass. Above, it extends to within two inches of the ensiform cartilage, and the lower margin was rather more than this distance from the navel. To the left it passed under the costal margin opposite the seventh and eighth cartilages. To the right it reached nearly, but not quite, to the costal margin. There was a distinct concavity above and a convexity below, and at no time was the mass below the level of a line joining the tips of the eleventh ribs. It was at times much more prominent than at others. On palpation, it was firm, gave an impression of solidity, was quite painless, of uniform resistance in all parts, and toward the left passed beneath the costal margin. To the right it terminated at a much higher position than is usual for the pyloric orifice. No movements were noticed in it, but the hand placed upon it occasionally felt distinct gurgling. Everywhere over it there was, on percussion, modified resonance. There were no nodules.

At no time during the patient's stay in the hospital was there vomiting, nor could it be said there was marked distaste for food. At first he took an ordinary diet, but, finding that it gave a good deal of distress, it was replaced by a liquid diet of egg albumin and milk, which agreed very well. Mentally he was very despondent.

Bismuth and soda were at first given an hour or so after eating, and they relieved promptly the sense of oppression. It was not thought worth while to distress him with attempts at lavage or test breakfasts. Patient left the hospital unimproved on the 19th.

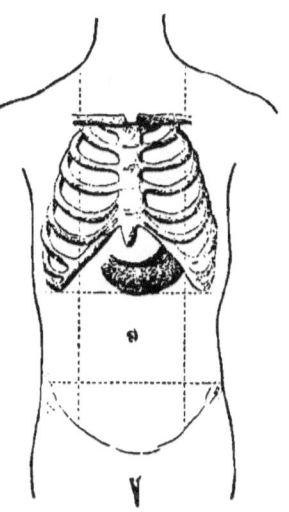

Fig. 18.—Area of the tumor in Case XXII.

Patient died September 28, 1892.

CASE XXIII.—*Unusually Large Cancer of the Stomach.*— Mrs. L., aged about fifty years, seen September 12, 1892. The patient has been a high-strung, nervous woman, and has not been in her usual health for the past two years, complaining chiefly of weakness and ill-defined nervous symptoms. I saw her a year ago for these symptoms, and at that time made an examination of the abdomen, which was negative. In June she was seen by a physician, who tells me that there was a lump on the left side which he thought was a floating kidney.

During the early summer she was under the care of a New York quack, who put her upon meat diet and hot water, under which treatment she appeared to improve. She went north to a watering place, and, though growing weaker and losing rapidly in weight, she kept up until August, when, on account of the swelling of the feet, she consented to go to bed. All this time she was chiefly on the meat diet, and apparently digested it very well, as she had no eructations and no vomiting. Subsequently she had a more varied diet and complained a good deal of distention and uneasiness after eating, and on several occasions had regurgitation of food. The lump in the left side had apparently increased in size, but caused her very little pain, except when it was rubbed by the masseuse.

The condition when I saw her was as follows: Profound emaciation, particularly marked in the face. The mind quite clear; the voice strong, and the grasp of the hand firm and good. In spite of great wasting, she did not look cachectic, and the color of the lips was good. Pulse 84, of fair volume; temperature normal. Tongue was red with a light furry coat. Her chief complaint was of uneasy feelings of distention after food, and of the weakness and prostration. The sleep was not, as a rule, disturbed, though she had been taking opium suppositories to allay the irregular pains in the side. She had no cough; no diarrhœa.

The abdomen was a little distended, contrasting with the extreme emaciation of the thorax. The upper zone was full and the skin over the left hypochondriac and umbilical regions reddened with applications, and these parts looked the most prominent. No peristaltic movements were noticed on inspection. On palpation, there was felt a large mass occupying the area shown in the annexed diagram, extending to the right 2·5 centimetres beyond the navel and the same distance below. The edge passed transversely to the left to a point four centimetres above the anterior superior spine; the edge could then be followed readily in the line of this spine to the point of the last rib. Above, it passed beneath the costal margin, and the upper line reached to within five centimetres of the ensiform cartilage. It felt superficial, firm, not tender; below and to the right the edge was unusually distinct, and just at the navel there was a slight depression. The hinder edge

NODULAR AND MASSIVE TUMORS OF THE STOMACH. 61

could be distinctly felt, and it did not pass deep into the renal region.

On bimanual palpation, the mass could be moved slightly. At the first examination there was no gurgling to be felt. On percussion, it was flat in the greater portion of its extent, but in the right fourth of the mass it was distinctly resonant.

There were no glandular enlargements. The blood examination was negative, with the exception of a very great increase in the blood plates.

I was not a little puzzled at first as to the nature of this tumor. The situation, the flatness, its superficial character, excluded definitely, it seemed, a movable kidney, which would not for a moment have been considered had I not been informed that a physician in whose judgment I have great confidence had in June pronounced this to be the condition present. The situation was suggestive, naturally, of an enlarged spleen ; the right edge seemed thin and there was an indistinct feeling of a notch, but the very superficial character, the absence of a definite notch or notches, and, above all, the resonance over one half of the tumor, seemed inconsistent with this view. A phantom tumor in a hysterical woman had also been suggested. The large size, the unusual situation, and the slight character o fthe

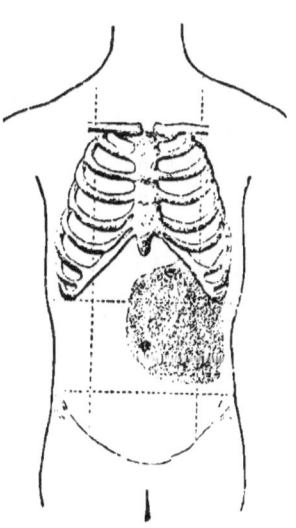

FIG. 19.—The tumor area in Case XXIII.

gastric symptoms did not favor gastric carcinoma, suggested, of course, by the profound emaciation and the existence of resonant tumor in the left hypochondrium.

The next few days, however, developed additional symptoms which made this view very much more likely. On the 12th she had taken six oysters and one on the 13th. On the 15th she had eructations of dark, very offensive material, and regurgitated one of the oysters in a condition of decomposition, but undigested. On the 16th she regurgitated the chopped meat in a similar condition of decomposition, and a second oyster which had been very slightly acted upon. The odor of the materials brought up was intensely offensive. She had also at this time slight diarrhœa. The tumor did not show any material changes, but the area of resonance seemed to vary somewhat, and, on drinking, gurgling could be distinctly heard over the mass, and sometimes with the hand upon the tumor the flatus could be felt.

September 19th.—For the first time the patient to-day had attacks of actual vomiting, the first at about six in the morning and the second at noon. On both occasions she brought up about half a pint of dark, bloody fluid of a most horribly offensive character, having a distinctly fæcal odor, as well as an odor of decomposition. In the material last vomited there were several grayish, sloughy masses the size of peas, which under the microscope did not show any definite structure. The patient after these attacks was much exhausted.

For a week after this there was vomiting and at intervals entire inability to take food, and occasionally vomiting of the same offensive material.

Died gradually of asthenia. There was no autopsy.

The extent of the tumor was due to infiltration of a very large area of the anterior wall and fundus of the organ. In all probability there was also extension to the omentum and to the colon. Evidently sloughing took place in the tumor mass, and, judging from the fæcal odor of the vomitus, perforation into the colon had occurred.

CASE XXIV. *Large, Massive Tumor in the Epigastric and Upper Umbilical Regions.*—Patrick C., harness maker, aged fifty-six years, admitted April 15th, complaining of weakness and a lump in the left side.

Family history is negative ; father died of accident ; mother, cause unknown, aged fifty-five years.

Patient has never been a very strong man ; was hurt when a lad by falling off a load of hay ; rheumatism in 1876. Has always

NODULAR AND MASSIVE TUMORS OF THE STOMACH. 63

been rather pale ; lived a sedentary life ; has not been a heavy drinker ; never had venereal disease. Within the past year he has lost between twenty and thirty pounds in weight.

Present illness began about a year ago with diarrhœa, which persisted for between eight and nine months. Sometimes he would have four or five stools in an hour, and often as many as twenty in a day. Never passed any blood ; sometimes would have none for two days. Was always better when he rested in bed. Thirteen weeks ago he left off work on account of the weakness, and during this time he has been rather inclined to constipation. He thinks he has become paler. One day about three months ago he felt a lump under the left ribs, and this he thinks has increased in size. His appetite has been variable ; lately it has improved somewhat. He has had no nausea and no vomiting ; no trouble in digesting his food. Once or twice has had slight nausea after taking milk. Has had no pain ; only a slight heavy feeling in the left side.

Present Condition.—Patient is pale, but not specially emaciated ; hair and beard gray ; conjunctivæ pearly white. Tongue lightly furred. Pulse, 96 ; vessel wall not sclerosed.

Abdomen a little prominent to left of navel ; throbbing of abdominal aorta marked. Subcutaneous veins not enlarged. On palpation, occupying the left hypochondriac, the left half of the epigastric, and the upper and left part of the umbilical regions, is a large flat mass. To the right it extends a little beyond the navel, and here the edge can be felt, and at the lower part the suspicion of a notch. The upper limit is ill-defined. At the costal margin it is felt to extend a short distance under the ribs. On deep inspiration the hand can be placed between the costal border and the tumor, which in this region has a distinctly rounded globular surface. Above and to the right it can be separated distinctly from the liver. During inspiration and on deep palpation no splenic margin can be detected in the normal position. On prolonged palpation of the tumor, no changes in consistence can be felt ; no gas is noticed to bubble through it.

Percussion.—In the middle line there is no dullness. There is resonance over the whole tumor. A modified, flat tympany is elicited in the left half of the epigastric region. The stomach tympany begins in the parasternal line at the upper margin of the seventh, and extends to two fingers' breadth above the navel. The artificial inflation of the stomach did not show any marked changes, and no gas was felt bubbling in the mass. The distention of colon with air made no change in it.

The mass is distinctly, though not very freely, movable. It can not be pushed back under the costal margin, and in fact can not be moved to the left so that its right margin is beyond the middle line. It does not extend deep into the renal region, palpation in which is normal. The tumor mass is not at all sensitive. The liver dullness is reduced; in the nipple line not more than a finger's breadth; in the axillary line no actual dullness. Though there is this extreme diminution in the area of liver dullness, the edge of the organ can be distinctly felt just below the costal margin. Although the edge of the spleen could not be felt under the left costal margin, there was at the first examination definite splenic dullness over the ninth and tenth ribs in the mid-axillary line.

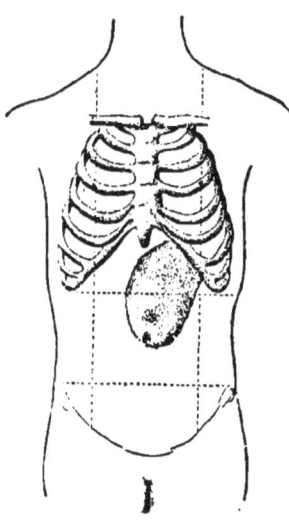

Fig. 20.—Outline of the tumor in Case XXIV.

The blood condition shows a profound secondary anæmia, and is not specially suggestive of a splenic form. The examination shows only thirty per cent. of hæmoglobin, a little over three million red blood-corpuscles to the cubic millimetre, and a little over nine thousand white corpuscles.

The urine is negative, pale, no abnormal deposits, no casts. He is constipated. Stools formed, and presented no special features.

A test breakfast showed no free hydrochloric acid.

Comments.—April 19, 1893. It has been suggested that this mass might possibly be a dislocated and fixed spleen, but the solidity and firmness, the rounded character of the mass, and the indefiniteness of the notches (supposed to be felt), were against this. Moreover, dislocated enlarged spleens are usually very mobile, and, most important of all, the splenic dullness was quite marked in the mid-axillary line, and the mass was everywhere resonant.

One of the most suggestive features of the case is the onset of the disease with diarrhœa. The situation of the mass, its fixity, and its size are against tumor of the colon. There are, on the other hand, instances of malignant disease of the small bowel in which the tumor mass has attained a very large size, and in which progressive emaciation, anæmia, and diarrhœa have been the main symptoms. The most remarkable case of this which I call to mind is one which I saw at the General Hospital, Montreal, with Dr. Molson, for whom I made the dissection. A man, aged forty-one years, was admitted on March 4, 1882, with swelling of the feet, vomiting, and constipation. For six months he had had pains and vomiting and more or less constipation, with loss of flesh. The patient had general anasarca and shortness of breath. The abdomen was full and large, and the examination was very difficult on account of the infiltration of the abdominal walls, and there was ascites.

The post-mortem showed a large mass occupying the left half of the abdomen, from the ribs to the crest of the ilium. It was firmly attached to the left kidney behind, and the colon and sigmoid flexure were at its left border. It was removed with the small intestine, and the tumor was found to involve eighteen inches of the jejunum, which tunneled the mass in a curved direction. The walls were in places from six to eight inches in thickness. The lumen was expanded, the mucosa still evident, presenting blunt valvulæ conniventes.

In Case XXIV the main portion of this tumor is in the epigastric region in the situation of the stomach. Though firm and solid, it was resonant, and, in the absence of definite features, the probabilities seemed to me that it was a large, massive tumor of the anterior wall and fundus. He died in October, with what symptoms we could not learn. There was no autopsy.

. Finally, let me sum up a few leading points for your guidance, based on the study of the cases we have had under consideration:

1. Though only a small section of stomach is available for palpation, a very large proportion of all tumors of the organ may be felt, owing in part to their greater frequency at the pyloric portion, and in part owing to the frequent depression of the organ. In every one of the twenty-four cases a tumor or induration was detected, and it is interesting to note that in the same period of time during which these cases were observed no instance came to autopsy with a tumor at the cardia or posterior wall.

2. In a considerable number of cases the dilated stomach itself forms a tumor in the abdomen, characterized by undulatory peristalsis, sometimes by a definite stomach contour. In ten cases of the series these features were distinct enough to render the diagnosis clear on inspection alone.

3. In a majority of cases no serious trouble is experienced in determining whether or not a tumor is in the stomach. Excessive mobility of a pyloric growth and extensive infiltrating masses in the epigastric region were the only conditions causing trouble in any of the cases of this series. The more systematic and thorough the examination, the less is the liability to error.

4. The character of the tumor is rarely in doubt. Large, nodular, and massive growths are invariably cancerous. At the pylorus it may be difficult to distinguish between cicatricial thickening about an ulcer, hypertrophic stenosis, and annular scirrhus. It may, in fact, be impossible to decide the question. The age, previous history, the general and local conditions—all have to be carefully taken into account, but, as in cases XIII, XIV, and XV, it may not be possible to reach a definite conclusion.

And, lastly, the very serious nature of tumors of the stomach may be gathered from the fact that, of the twenty-four patients, eight have already died.

LECTURE III.

TUMORS OF THE LIVER.

LET me remind·you of two anatomical points: First, that the liver extends entirely across the upper portion of the abdominal cavity, so that tumors may project to either side of the middle line in front, more rarely into the left pleura behind; and secondly, that in women the organ is frequently so dislocated that a large part of the convexity is in contact with the anterior abdominal wall; moreover, the anterior margin may be irregular from the projection of tongue-like portions, to which special attention will be directed in considering tumors of the gall-bladder.

Tumors of the liver are common, and, as a rule, their nature is not difficult to recognize. I shall not here refer to the simple enlargements of the organ in hypertrophic cirrhosis, in amyloid and fatty degenerations, or to the cases of uniform increase in volume met with in cancer and abscess. I shall call your attention only to those in which a prominent nodular mass or swelling—a tumor—was detected, and the nature of which had to be decided.

The usual causes are cancer, abscess, syphilis, hydatids, and occasionally tuberculosis. The tumors in connection with the gall-bladder I shall consider separately. Under certain circumstances the liver itself may form a tumor-like structure. The cases which have come before me for diagnosis in the past twelve months are distributed as follows: The liver itself, one; abscess, four; syphilis, two; cancer, four.

68   THE DIAGNOSIS OF ABDOMINAL TUMORS.

I. TUMOR FORMED BY THE LIVER ITSELF.—I show you here a little patient (Case XXV) in the upper part of whose abdomen you can see, even from a distance, a prominent tumor, which pulsates actively at the rate

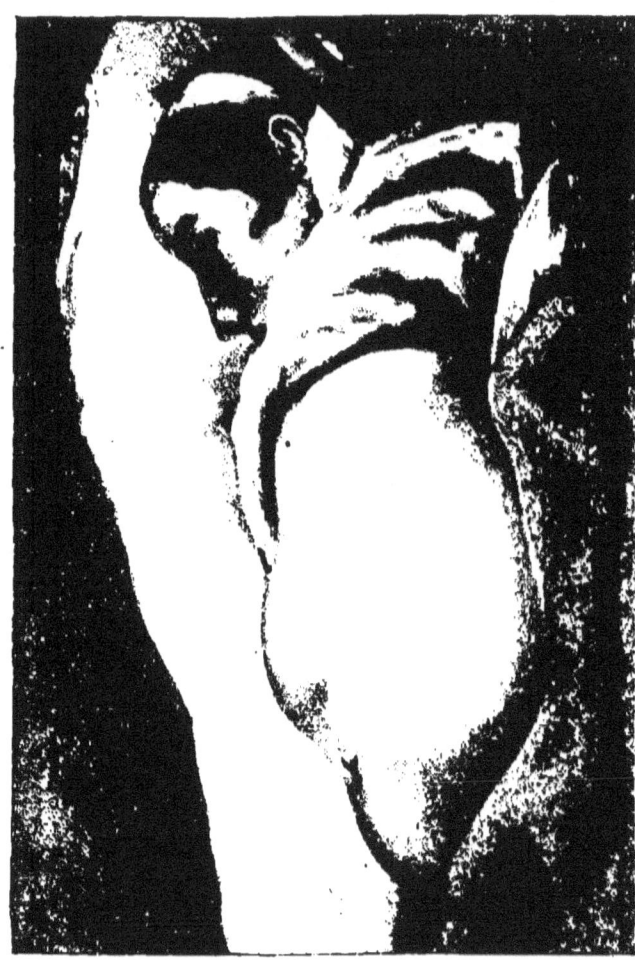

FIG. 21.—Pulsating tumor in the epigastric region consisting of the rounded and contracted liver.

of ninety per minute, lifting the skin in the epigastric region. This case has been under our care on and off for the past two years. She is thirteen years of age, and has an old mitral-valve lesion from rheumatism, with enormous enlargement of the heart. The apex beat is, as you see, far out in the sixth and seventh interspaces. The præcordia is very prominent, and there are signs indicating that the pericardium is adherent. During the past eighteen months ascites has constantly recurred, so that she is now tapped once a week, yesterday for the seventy-first time. When the abdomen is distended nothing is noticed, but after the fluid is withdrawn this remarkable tumor-like mass appears in the epigastrium (Fig. 21). On palpation it is smooth, with a rounded edge, descends with inspiration, and expands visibly; and under the fingers, during the cardiac systole, it can be traced to the right, where at about the nipple line it passes beneath the costal margin. As it pulsates there can be felt, particularly at these periods, a to-and-fro peritoneal friction rub. The pulsation is expansile, and with the fingers of the left hand beneath the costal margin in the nipple line, and the right hand over the prominent mass, the whole structure can be felt to expand with each contraction of the heart. The situation, the shape, and the character of the pulsation leave no doubt whatever that this is a pulsating liver, a not very uncommon condition in chronic mitral disease, when the tricuspid becomes insufficient and allows each systole of the right ventricle to be communicated through the right auricle directly to the column of blood in the hepatic veins. The deformity of the liver, its cakelike shape, and rounded edge are caused, I believe, by a perihepatitis, possibly a direct extension from the pericardium, associated with which there is a chronic proliferative peritonitis. The recurring ascites is due partly to the contraction of the liver by the perihepatitis, partly to the chronic peritonitis. A case with almost identical features was for a long time under my observation at the University Hospital, Philadelphia.

In two other conditions—neither of which, however,

has been before us this year—the liver itself may form a tumor and cause no little difficulty in diagnosis: first, the so-called floating liver, which is most commonly met with in women (though sometimes met with in men, as in a recent case reported by Kreider, of Springfield, Ill.), and is a feature of enteroptosis; and second, the cases of great shrinkage and deformity of the liver in syphilis, in which the whole organ may be converted into a cluster of irregular nodular masses, held together by fibrous tissue, a condition to which the term *botryoid* has been given from its resemblance to a bunch of grapes.

II. ABSCESS OF THE LIVER.—Unfortunately for the victims of this serious disease, a prominent tumor mass is only occasionally present. Of nine cases of abscess of the liver seen since the first of September of last year, of which seven were in the hospital (hospital numbers 5876, 6109, 6745, 7679, 7687, 7738, and 8001), four presented a prominent tumor, the nature of which came up for discussion.

CASE XXVI. *Abscess of the Liver; Prominent Tumor; Incision; Recovery.*—Dr. X., aged sixty years, admitted September 6, 1892, complaining of weakness and of a painful tumor in the side. Family history good. The patient has been a healthy man and has had very few illnesses, the only severe one being typhoid fever, in 1863. The present trouble dates from April of this year, when he began to have pain in the right side, fever, and chilly sensations. The temperature sometimes rose to 103°, and he had a sense of distention and fullness in the right side, but there was no bulging, as at present. No history can be obtained of any local disease in the gastro-intestinal tract. He is positive that there had been no diarrhœa. After a month, during the greater part of which time he was in bed, the fever disappeared, but the pain and fullness in the right side persisted. Toward the end of June he noticed that there was a prominence below the right costal margin which has steadily increased. He has lost much in weight—from 250 to 185 pounds. Since June he has had at times slight fever, but

TUMORS OF THE LIVER. 71

no chills, and only occasional sweating. The bowels have been irregular, and he has had to take purgatives. There has never at any time been jaundice.

Present Condition.—Fairly well nourished, a little pale, but

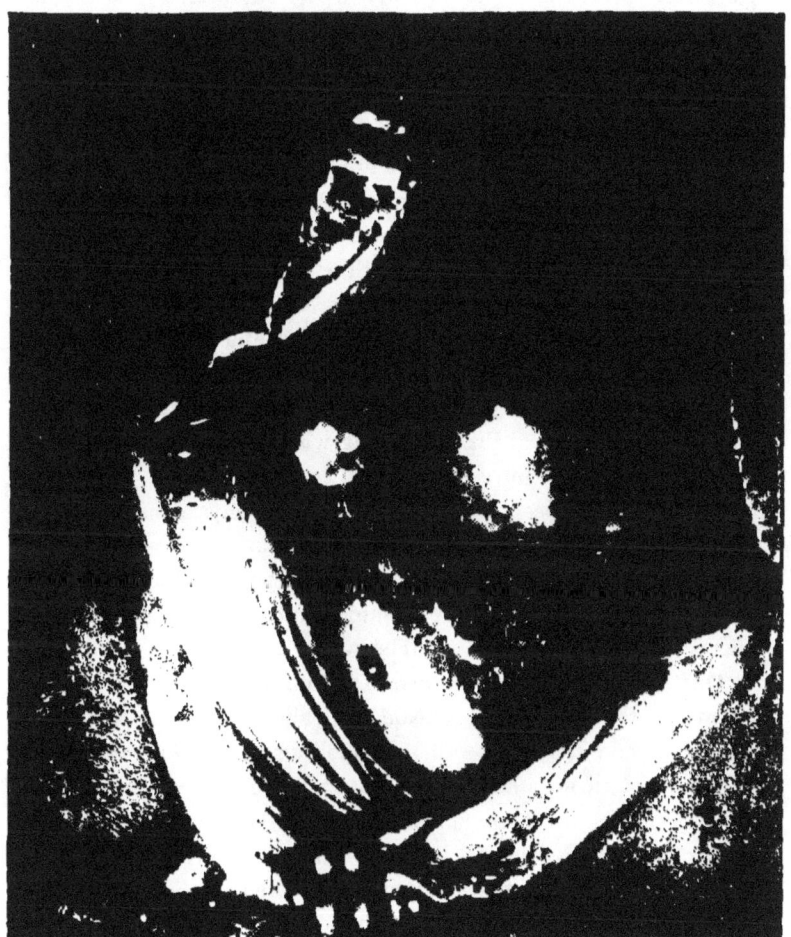

Fig. 22.—Abscess of the right lobe of the liver forming a prominent tumor (Case XXVI).

neither emaciated nor cachectic. Temperature last evening 101°; pulse 104, occasionally intermits. The radials are sclerotic. The tongue is red, not dry.

Abdomen.—A large tumor mass fills up the right half, and is strikingly prominent, as indicated in the figure from a photograph

(Fig. 22). Below, it reaches the transverse umbilical line; above, it passes beneath the costal margin. To the left the swelling begins at the middle line. The skin over it is glossy and a little reddened. The respiratory movements of the abdomen are slight. On palpation, the entire right side above the transverse iliac line is occupied by a solid mass which is resistant except at the most prominent point, where it is soft and fluctuating. It is nowhere painful on pressure. Below, no definite sharp edge can be felt. Above, it is continuous with and not to be separated from the liver. Behind, it occupies the entire flank and can be felt on deep pressure below the eleventh rib. On bimanual palpation it is fixed, not mobile. Percussion gives a flat note everywhere over the tumor. Liver dullness begins at the seventh rib and is continuous with that of the tumor mass. In the parasternal line there is a slight resonance between the margin of the ribs and the tumor. The spleen is not palpable; area of dullness not increased.

The cervical and inguinal glands are not enlarged. There is a soft systolic murmur at the apex; otherwise the examination of the thoracic organs is negative.

As doubt had been expressed by several physicians who had examined Dr. X. as to the nature of the tumor, an exploratory aspiration was made in the most prominent portion and a grayish-red pus withdrawn, which on examination contained much molecular *débris*, pus cells, fatty crystals, but no amœbæ.

On the 10th, under chloroform, Dr. Finney made an oblique incision over the tumor and opened a large abscess cavity in the liver, removing more than a litre of foul-smelling pus, darkish in color; on examination, no amœbæ were found. The walls of the abscess cavity, as felt by the finger, extended beyond the middle line and upward out of reach beneath the ribs. They were everywhere hard and firm. The patient reacted well after the operation ; the temperature fell, and on the sixth day he was wheeling himself about the ward in a chair.

Patient left the hospital September 28th, and when last heard from, six months after the operation, remained well.

CASE XXVII. *Abscess of the Liver; Prominent Tumor in the Epigastrium.*—Mr. W., merchant, aged fifty-four years, seen September 16, 1893, with Dr. Opie. Patient was a healthy man until

July, 1892. Has had the ordinary diseases, and syphilis when a young man. Twenty or more years ago he had diarrhœa for some time.

He dates his present illness from July, 1892, when without any vomiting or diarrhœa he began to have pains in the abdomen, which persisted with great severity for about two weeks. At this time he could not straighten himself without very great pain. The abdomen was swollen; he had no jaundice. He improved somewhat and went away for about six weeks and gained a great deal in weight. In September, on his return, he again had the severe pain and oppression in breathing. He was at this time under the care of Dr. Opie and Dr. Chambers, who state that the liver was greatly enlarged. Abscess was suspected, and an operation was suggested but declined. The swelling was not so marked as it is at present. He does not appear to have had any chills, sweats, or indeed much fever. There never has been jaundice, and it was subsequently suggested that the enlargement of the liver was due to syphilitic disease. He lost a great deal of flesh during this illness—as much, he thinks, as fifty pounds. Throughout the early part of this year he was very much better and gained about twenty-five pounds in weight. He went away in March, but was not at all benefited by the change. He lost strength and flesh, and lately has had a great deal of dragging pain in the side, particularly if he attempts to lie on the left side. He has had no diarrhœa and no vomiting.

The patient is a large-framed man, with sallow complexion, looks ill, and is decidedly emaciated. He was sitting up; no swelling of the feet; conjunctivæ pale but not jaundiced. The pulse is 96, tension not increased; he has no fever.

The abdomen is enlarged and a prominent tumor fills the epigastric region and extends toward the left hypochondrium. The skin over it is glistening, dry, and abraded from counter-irritation. The superficial veins are not specially enlarged, except the left mammary, which is prominent. No enlargement of the superficial glands. On palpation, the abdomen is soft and natural until just at the level of the navel. Here the edge of the liver can be distinctly felt. To the left the edge passes obliquely and can be felt to pass under the costal margin at the ninth cartilage. To the right the edge passes obliquely upward and can be felt at the costal margin at about the tenth rib. Occupying the right epigastric region there is a prominent flat projection which causes a distinct asymmetry in this region. It is not painful on pressure. It is soft and boggy, but there is no definite fluctuation. Percussion shows the liver dullness to be greatly increased. Above, it extends nearly to the lower margin of the fourth rib, crosses the sternum opposite the cartilage of the fifth rib, and is continuous with the cardiac dullness. There is a vertical liver dullness from the fourth interspace to the level of the navel in the parasternal line. Behind, the dullness reaches very high, almost to the angle of the scapula.

The spleen is not enlarged. The heart is a little pushed up; the sounds are clear. The examination of the lungs is negative.

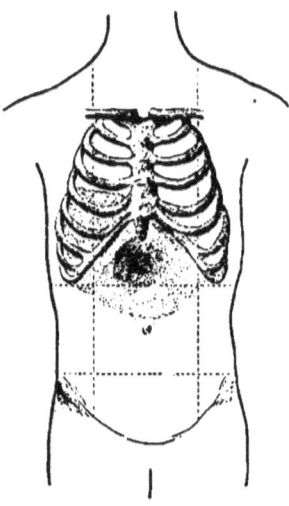

Fig. 23.—Outline of the liver and situation of the tumor in Case XXVII.

The diagnosis of the condition was not at all easy. The progressive emaciation and the enormously enlarged liver and somewhat irregular outline suggested, of course, cancer, against which, however, was the notable fact that he had improved so much after a very severe attack last year, in which the liver was enlarged. The prominent hemispherical mass in the right epigastric region was suggestive of abscess. Though he had had no chills and no fever, and although in the history not one of the usual ætiological factors preceding abscess of the liver was present, the sallow cachexia, the dragging pains on attempting to lie on the left side, and the prominent doughy tumor of the liver, made an exploratory examination advisable. An aspirator needle was thrust deeply into the most prominent part of the tumor and immediately a grayish and subsequently a reddish-brown pus flowed out freely.

The patient was removed to the City Hospital, where on September 21st Dr. Chambers opened the abscess and removed a gallon and a half of reddish, thin pus, which microscopically was made up of a granular *débris*, very few pus cells retaining their contour; no amœbæ. The patient rallied from the operation, but sank and died in about ten days.

CASE XXVIII. *Abscess of the Liver; Tumor in the Right Epigastric Region; Rupture into the Lung.*—Simon G., aged twenty-seven years, admitted July 11, 1893, complaining of hiccough and pain below the ribs on the right side. He has always been healthy until five years ago, when he was laid up in the Hebrew Hospital for six weeks with a severe cough. His habits are good and he has not had venereal disease.

Three weeks ago his present illness began with irregular cramps in the hepatic region. He had hiccough for nearly a week, day and night, and this was in reality his most distressing symptom. He has vomited several times; lost his appetite and has only been able to take milk and whisky. The bowels have been constipated and he had never had any severe diarrhœa. No cough; no expectoration. On July 1st he had a very severe chill in which he shook for an hour, followed by fever and sweating. The following day, July 2d, he felt better. On the 3d he had a second chill at eight o'clock in the morning, and shook for an hour. On the 4th he had another chill. He then went into the Norfolk Hospital, where he was given a great deal of quinine. He has sweated a good deal; has not noticed that he has become at all yellow. He has lost somewhat in weight. At present he complains of the incessant hiccough and of pain in the region of the liver.

Present Condition.—Patient is fairly well nourished; face a little emaciated; conjunctivæ slightly tinged. The skin is not jaundiced. Tongue has a light white fur. Temperature has not been above 99.5° since his admission; pulse, 84; tension normal.

On inspection, the thorax on the right side looks a little fuller than the left, particularly behind in the infrascapular region. The left intercostal grooves are faintly visible; the right not at all. The apex beat of the heart is not visible; the percussion is clear on the left side; on the right side in front it is clear to the upper border of the sixth in the nipple line, and to the upper border of the fifth in the midaxillary line. Behind there is defective resonance at the angle of the scapula, shading quickly into dullness. The respiratory murmur is heard well on both sides. It is feeble in the infrascapular region on the right side; no râles; no friction. The abdomen looks natural, but there is a marked fullness on the right side in the hypochondriac region, and the groove below the costal margin is completely obliterated. In this region, occupying a space thirteen centimetres in diameter, there is a very definite prominence. On palpation, this is resistant and tender, and the skin at the costal edge seems a little infiltrated. The liver margin can be felt reaching nearly to the level of the navel and the edge is rounded. To the right it passes under the costal margin at the tip of the tenth rib. The left lobe can be felt filling the upper epigastric region. The liver dullness extends high in the axillary region, reaches the upper border of the fifth, and there is here eighteen centimetres of vertical dullness.

The edge of the spleen is distinctly palpable. The superficial lymph glands are not enlarged and the examination of the stomach and intestines is negative. A rectal tube was passed, but no amœbæ were found in the mucus obtained. The blood examination was negative.

I aspirated in the parasternal line at the most prominent point of the tumor, but obtained no pus.

I did not see this patient again, but abstract the following from Dr. Thayer's notes : Although no pus was obtained on the first aspiration, there seemed to be no question as to the correctness of the diagnosis. Chills occurred on July 14th, 17th, 19th, and 23d, and on August 1st, 3d, and 4th. The patient was urged to have an operation, but declined. The tumor mass remained prominent, but no definite fluctuation developed. On August 20th the patient suddenly began to cough, calling for the nurse and saying that he felt something had burst inside him. He expectorated several spit-cupfuls within a short time of a dirty, yellowish-green pus. The odor was not offensive ; microscopically, it showed degenerated pus

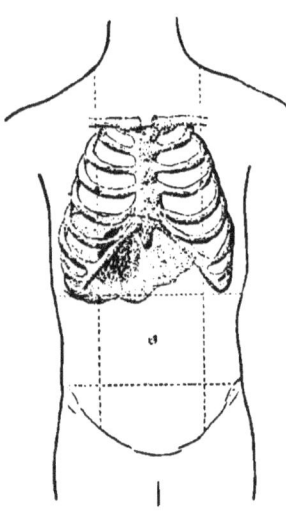

Fig. 24.—Outline of the liver and situation of the tumor in Case XXVIII.

cells; no amœbæ. An interesting feature is that on the following day the prominence in the right epigastric region had almost disappeared. The physical signs at the right back had not changed. Over the area of dullness the respiratory murmur was simply enfeebled, and there were a few fine râles on coughing. The temperature range was from subnormal to 102°; once only, after a chill, 103·3°. He was repeatedly advised to have an operation, but declined, saying that he would sooner take his chances. The expectoration of purulent matter continued, but in diminishing quantities; thus on the 23d it was one hundred and twenty cubic centimetres; by the 26th, thirty cubic centimetres; and on the 28th, forty cubic centimetres. On August 1st it had fallen to only ten cubic centimetres; on the 3d he had none, and on the 6th only ten cubic centimetres. No amœbæ were found, only pus cells, most of them in a condition of disintegration. On August 8th he was taken home by his brother, the condition not having materially improved, and the temperature still ranging from 98° to 102°, occasionally to 103°. The prominent mass on the right side never reappeared. On discharge, the liver flatness began in the fifth interspace in the nipple line; the border could be felt five centimetres below the costal margin. In the midaxilla there were thirteen and a half centimetres of vertical dullness. In the median line from the upper limit of flatness to the lower border, determined by palpation, was fourteen centimetres.

CASE XXIX. *Acute Dysentery; Abscess of the Left Lobe of the Liver; Tumors in the Left Epigastric Region; Incision; Death.*—Raphael F., tailor, aged twenty-seven years, admitted August 22, 1893. His family history is good. Patient is a Russian, and has been in this country only nine years. Has been always healthy and strong.

Present illness began abruptly about a month ago with an attack of vomiting and purging; had six or eight stools the first day. The next day he was rather better and was pretty well for two or three days. Then he again had vomiting and much purging. Evidently the attack was one of acute dysentery, as he had numerous stools containing slime and blood, passed with much straining and tenesmus. He does not think he had at this time any fever. He has lived almost entirely on milk. The diarrhœa has continued ever since. On admission, he was thin, lips and mucous membranes of good color; tongue covered with a thick, white coating. The temperature was 97°. The examination of the thoracic organs was negative. The abdomen was retracted; patient complained of pain on palpation over the cæcal region. The liver dullness began in the sixth interspace in the nipple line and extended to the costal border. The edge was not palpable. He had three stools within the first twenty-four hours, brownish in color, fluid, and containing small strings of mucus. In the mucus obtained by passing the rectal tube numbers of actively moving amœbæ were found. He was ordered large, high injections of sulphate of quinine—one to two thousand—and a milk diet. During the first week in hospital there was no special change. The passages were from four to seven in twenty-four hours. The temperature was never above 99°, and he seemed to be doing well. For the next two weeks also he seemed to be very comfortable. The temperature between August 29th and September 11th was not above normal. The dysenteric symptoms had improved very much, and from September 9th to the 13th he had only had one or two movements a day. On September 14th he made, for the first time, complaint of a pain in the epigastric region, and there had been for two days slight fever, the temperature rising to 100°. There was a little sensitiveness over the liver in the middle line, but the organ did not appear to be enlarged. On September 17th there was noticed for the first time a rounded prominence occupying the left half of the epigastric region, which moved up and down with respiration and which pulsated with each heart beat. It extended from the costal margin in the parasternal line to a little beyond the middle line. It was rounded, smooth, firm, but elastic, and did not appear to fluctuate. The pulsation was very marked and seemed almost expansile. On placing the patient, however, in the knee-elbow position the pulsation en-

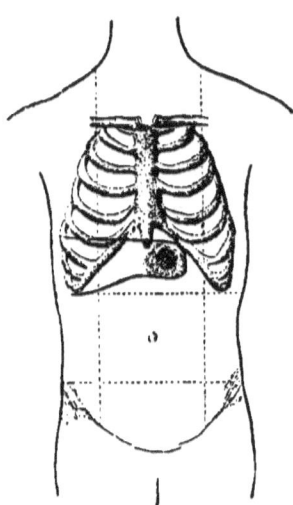

FIG. 25.—Outline of the liver and situation of the tumor in Case XXIX.

tirely ceased; no bruit was heard over it. The situation of the tumor is indicated in the annexed diagram. The liver dullness began at the sixth rib and extended just below the costal margin in the nipple line. In the parasternal line it did not begin until the seventh rib, and in the middle line the dullness was about the middle of the ensiform cartilage and was then continuous with the tumor mass on the left side. In the midaxillary line the dullness began at the seventh rib and extended to the costal border. The splenic dullness could not be obtained, nor could the margin be felt. There was no increase in the hepatic dullness behind.

On the 18th the temperature rose to 101·2°, the highest point reached, and the patient had no chills, but was sweating profusely. The tumor was more prominent; was very tender, but there were no signs of fluctuation. The diagnosis of abscess of the liver was made and the case was transferred to the surgical wards. Dr. Finney operated, found the peritonæum adherent to the liver, and opened a superficial abscess in the left lobe. About seventy-five cubic centimetres of thick, yellowish-green pus were evacuated, in which amœbæ were present. The patient stood the operation very well, but he became progressively weaker. The temperature never rose above 101·5. He died on the 24th.

The situation of the palpable tumor in liver abscess is well illustrated by these four patients. In Case XXVI it was in the right lumbar and right side of the umbilical regions, entirely below the costal margin. In Case XXVIII it was in the right hypochrondriac region. In Case XXVII it was median, projecting prominently midway between the navel and ensiform cartilage; and in Case XXIX it was entirely to the left of the middle line in the upper quadrant of the epigastric region.

In the diagnosis of liver abscess you must take into consideration the following points:

Antecedent Conditions.—Dysentery is in this latitude by far the most common, though it was only present in one of the four cases I have narrated. Of the nine cases, however, under observation during the past year, dysentery occurred in six cases. There may, however, be no recognizable cause, as in Cases XXVI, XXVII, and XXVIII. Bear in mind that when the patient comes before you with chills and fever, a sallow cachexia, and an enlarged, tender liver, the dysenteric symptoms may have entirely disappeared, or there may be only at intervals slight recurrences. Under these circumstances a catheter or the long rectal tube passed into the bowel may bring away portions of mucus containing the amœbæ, associated with the severer form of dysentery.

Toxic Features.—Irregular fever, chills, and sweats are rarely absent. The sallow tint of the skin, the progressive anæmia, and the paroxysms of intermittent fever lead very frequently to a diagnosis of malaria.

Local Symptoms.—Increase in the size of the liver and tenderness on pressure are the most important. The enlargement is most frequently of the right lobe, but the whole organ may be greatly increased in size and extend below the navel. When the abscess is in the right lobe the enlargement may be chiefly behind, ascending high into the right pleura. Prominent bulging of the lower portion of the right side of the thorax is extremely common.

And lastly, and what interests us here especially, a tumor mass may develop beneath the right costal margin or in the epigastric region. The tumor is usually (always when recent) tender; often develops with rapidity. The rapid increase in size with tenderness, however, is not to be relied on as characteristic, as I will mention to you in the fourth of the cases in the cancer series this was a very marked feature. Fluctuation may be obtained readily when the tumor mass becomes superficial. The tumor may persist, as in Case XXVI, for months without very much change. With or without the presence of a tumor, when liver abscess is suspected, the long aspirator needle should be freely used.

III. Syphilis of the Liver.—Of four cases diagnosticated during life as syphilis of the liver, two presented definite tumors. Diffuse syphilitic hepatitis does not produce a tumor, but gummata, either in the inherited or acquired disease, may form tumors in two stages: first, when fresh and developing, constituting nodular masses of large size, which may persist for months; and, second, gummata which have undergone cicatricial contraction and healing may so fissure and divide the liver by bands of connective tissue that an extremely nodular, irregular mass may oc-

Fig. 26.—Showing the extreme irregularity of a syphilitic liver.

cupy the right hypochondrium. Of the four patients this year, two died, but in neither of them were tumors felt. I show you here a photograph of the liver of one of them (Fig. 26), which will give you an idea of the extraordinary subdivision of the organ, an extreme grade of which forms the so-called "botryoid" liver, in which globular masses of normal tissue are held together by fibrous bands. The other case also, which came to autopsy, had an extremely irregular liver, and was of exceptional interest, inasmuch as the recurring ascites, for which she had been tapped repeatedly, disappeared entirely under iodide of potassium, as did also nodes on her shins. In the following cases definite tumor masses were present and the correctness of the diagnosis in both instances was in a measure borne out by the therapeutic test. In anomalous tumors of the liver it is well to bear in mind that gummata may form flat or nodular masses in the epigastric region * which may persist for a long time, and which may, under treatment, disappear as satisfactorily as gummata of the long bones

---

* A case illustrating an error in diagnosis is that of hospital number 5234, Joshua M., aged fifty-four, admitted May 16, 1892, with swelling of the abdomen and an illness of nearly a year's duration. The patient was a large, powerfully built man; had always enjoyed good health, and denied venereal disease. For nearly a year he had had trouble in the abdomen, and had twice been slightly jaundiced. The legs had been swollen, and he had had shortness of breath. The examination showed an enormously enlarged liver. The whole of the upper part of the abdomen was filled with a hard, irregular, nodular mass, corresponding to the greatly enlarged liver. The lower border was felt midway between the umbilicus and the pubes. There were prominent bosses on the surface of the liver, and from its great size and irregularity there seemed to be no question as to the correctness of the diagnosis of secondary cancer of the organ. No primary disease could be determined, and there was decided hyperacidity of the gastric juice. The patient remained in the hospital for a month and gained slightly in weight, but the liver developed still further, and the irregularity on the surface was more marked. An aspirator needle was thrust in, but nothing but blood obtained. Naturally enough the diagnosis was entered as cancer of the liver. Dr. F. T. King, of Washington, under date of February 10, 1894, writes that he has Joshua M. under his care at present. He has a well-marked syphilitic skin eruption, and the liver, still enlarged, extends nearly to the pelvis. The time element in this case, I should think, definitely excludes cancer, while the syphilitic rash on the skin is suggestive in the highest degree that the whole trouble is specific. On the death of this patient in the spring of 1894, Dr. Lamb, who made the autopsy, stated that it was cancer of the liver, an instance of unusually long duration.

or of the testes. One of the first private patients who applied at the hospital was a young man aged about twenty-eight years, who presented just below the ensiform a flat tumor mass evidently attached to the liver, the nature of which had been very much discussed. He was sent to me for an opinion as to the advisability of a laparotomy. A positive history of syphilis was obtained and he was urged to have a thorough course of treatment. I did not see him again for nearly a year, when, to my astonishment, the tumor had practically disappeared.

CASE XXX. *Prominent Tumor Mass in the Epigastrium; Disappearance in Four Months under the Use of Iodide of Potassium.*—John C., aged thirteen years, seen November 11th with Dr. Thayer. He had been in hospital during my absence in the summer with a tumor connected with the liver. The notes made at the time are as follows: Admitted July 13, 1892, complaining of pain in the right hypochondrium.

The father is living and is paralyzed; mother is living and well; two sisters and one brother died young; has three brothers and four sisters living.

As a child he was well, with the exception of enlarged glands, which, at about his fourth year, appeared on the right side of the neck and discharged for nearly three years.

His present illness began about three months ago with pain in the right side, which has continued, and two weeks ago became much worse; the abdomen became swollen, particularly in the upper zone and to the right side. The feet have never been swollen nor has he had any swelling of the face. Since his illness began there has been progressive loss of weight, and he frequently sweats profusely at night.

Present Condition.—Poorly nourished and not very well developed lad. Lips and mucous membranes of a good color. Skin clear, not jaundiced. Scars on the neck just below the right ear and one on the episternal notch. The cervical glands are not enlarged; the inguinal glands are only just palpable. The epitrochlear glands are enlarged; there are no nodes, though about the middle of the right tibia there is a slight roughening. There are rhagades at the angles of the mouth ; the corneæ are clear ; the upper central incisors are well formed. Examination of the blood is negative.

The thorax is somewhat expanded in the lower part, particularly on the right side. Examination of the lungs and of the heart is negative.

The abdomen is prominent in epigastric and right hypochondriac regions. The hepatic flatness begins in the sixth interspace in the nipple line and extends two fingers' breadth below the costal margin. The epigastrium is filled with a firm, hard mass, which appears to be the left lobe of the liver or a mass continuous with it. It extends in the median line within two centimetres of the umbilicus, and a slight notch can be felt. The surface is somewhat irregular and a distinct rounded nodule can be felt a little above and to the right of the navel. In the nipple line the edge of the liver is difficult to feel, as the abdominal walls are rigid. The splenic flatness begins at the seventh rib and is continuous with that of the mass in the epigastrium. The edge of the spleen is not palpable.

The urine was clear, lemon-colored ; no albumin ; no tube casts. He had slight fever ; temperature on admission 100·5° F., and every day for a week it rose to 100° F. The patient was placed upon iodide of potassium, five-grain doses three times a day, increasing rapidly until he took a drachm three times a day. He improved in weight and gained seven pounds within a month. The mass in the epigastrium gradually disappeared.

The condition to-day, November 11th, is as follows :

The boy has grown somewhat, though he looks thin. The tongue is clean ; the lips and mucous membranes are of good color. He has no cough.

He certainly has not a luetic facies, though the scars at the angle of the mouth are suggestive. The edge of the liver does not appear to be below the costal border in the nipple line ; flatness begins at the sixth rib in the nipple line. Palpation in the epigastric region is negative until the extremity of the angle is reached, and here, just below the tip of the ensiform cartilage, can be felt a hard, firm ridge with, in the right xiphoid angle, a little prominent projection the size of a walnut. Just to the left of the tip

TUMORS OF THE LIVER.

of the ensiform cartilage a second prominent but smaller elevation can be felt. The one to the left, when he draws a deep breath, can be seen distinctly in the descent of the liver as a slight prominence beneath the skin. The edge of the liver comes about a finger's breadth below the tip of the ensiform cartilage and fills up the entire left costo-xiphoid angle.

This case is similar in some respects to one reported from my clinic at the University Hospital by Dr. A. C. Wood,* the history of which is so interesting that I abstract it here: "He had an eruption on the skin when six months old. Did not have snuffles. Has always been robust. For several months back his mother has noticed his abdomen becoming more and more prominent. The patient has lost flesh of late. Has had no jaundice. The boy, aged thirteen, is fairly well grown, well nourished, looks a little pale; abdomen prominent,

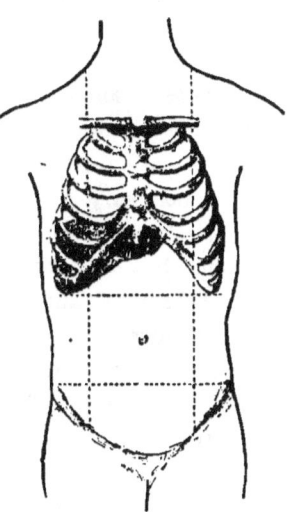

FIG. 27.—The situation of the tumor nodules in Case XXX.

and on the right side between the navel and the costal margin there is a distinct hemispherical swelling about two inches in diameter. The tumor descends slightly on inspiration. The superficial abdominal veins are not especially dilated. The upper teeth are not good, but are not notched. He has not a syphilitic facies.

"On palpation, the lower zone of the abdomen is soft. In the left hypochondriac region the edge of the spleen is distinctly felt, with its notch at least two fingers' breadth below the costal margin. Toward the right hypochondrium a firm, solid mass is felt, the edge of which is ill-defined below and to the right. Above, it seems to pass directly beneath the costal margin in the position of the liver. The tumor is painless. The liver dullness begins in the midsternal line at the level of the sixth costal cartilage, in the nipple line at the upper border of the fifth rib, and is directly con-

---

* *University Medical Magazine*, vol. ii.

tinuous with that of the tumor and reaches to within a finger's breadth of the navel. In the axillary line there is dullness from the upper border of the eighth to the lower border of the tenth rib. The left lobe of the liver does not appear to be enlarged. The splenic dullness begins at the upper border of the eighth rib and extends two fingers' breadth below the costal border.

"Fowler's solution was prescribed, four minims three times daily, to be increased by one drop each week. This was taken at intervals until the middle of January, about five months, when it became necessary to discontinue it on account of nausea and vomiting. During this time the color of the face had improved, the tumor had enlarged, and, on deep inspiration, the margin now reached the navel, and it was rough on the surface.

"In October, 1888, the patient complained of pain in both tibiæ. The pain was thought to be periosteal, and five grains of iodide of potassium were ordered to be taken three times a day. When seen again, two months later, the tumor was more nodular, the spleen had increased in size, extending three inches below the costal margin. The general condition of the patient remained good. He was now given one grain of calomel three times daily, and instructed at the end of three weeks to intermit for a week. This was continued with similar intermissions for about six months, with gradual improvement in appetite, color, and general health."

After leaving Philadelphia the boy came under the care of Dr. A. C. Wood, who reported that he returned to the hospital in February, 1890, and the mother thought there had been a slow improvement. There was, however, on the forehead a small region of necrosis of the frontal bone about the size of a ten cent piece, which had followed a definite node. The interesting thing is that the hemispherical swelling in the hepatic region, which was so striking in this case, had practically disappeared. I demonstrated this case on several occasions before my class in the session of 1887-'88 and of 1888-'89, and on each occasion the hemispherical swelling on the right side, between the navel and the costal margin, was unusually distinct—so much so that it could readily be seen by the students from the most distant benches of the amphitheater. I confess that until October, 1888, when there were pains of the tibiæ, I did not think the case syphilitic, but regarded it as an anomalous tumor of the organ. In October, 1888, however, he was given iodide of potassium and subsequently calomel. Its syphilitic nature seems to be definitely established by the development of the gummatous tumor on the forehead, which subsequently broke down and left a patch of necrosis.

CASE XXXI. *Syphilis of the Liver; Convulsions; Right Hemiplegia; Irregular Tumor over the Left Lobe.*—On May 3d J. M., aged forty-seven years, returned by appointment to report on his condition. I had not seen him since the 25th of September of last year. The case is one of a good deal of interest with reference to the diagnosis and treatment of syphilis of the liver. I saw him first in March, 1892.

The patient is a stout man, well built, but looks ten years older than the age he gives. He is a traveling salesman, and, in response to an inquiry as to his habits, gave the characteristic reply that "he took his luck on the road." He had syphilis in 1866, and was treated for some time. For eighteen months or more he had been complaining of dyspepsia and irregular pains in the abdomen. In December, 1890, he had vomiting, and last March, just a year ago, the pains were very severe—so much so that he had to have a hypodermic injection of morphine. He had no jaundice after this attack. He went to the country, stayed until June, and improved a great deal, but he had there another severe attack of pain in the abdomen. Through the summer he lost thirty-five pounds in weight. In the autumn he had an attack of jaundice which lasted for nearly two months and gradually disappeared. This jaundice set in with pains, which were very severe. Two months ago, while sitting in his office, he fell, lost consciousness, had a convulsive seizure, followed by left hemiplegia. Gradually the power returned. He had another convulsive seizure, with loss of consciousness, a week ago, not followed by paralysis.

When seen, March 24th, he was well nourished, not jaundiced. The point of special note was the examination of the abdomen. The panniculus was thick; the liver was enlarged. The right lobe felt somewhat irregular, but there were no definite nodules. The gall-bladder was not palpable. In the left hypochondriac region, emerging beneath the costal border, was a flat tumor mass, which extended in the parasternal line nearly to the level of the navel.

It was sensitive, firm, felt about the size of the palm of the hand, and descended with inspiration. It was superficial, no definite edge could be felt, and to the right it could be separated clearly from the edge of the liver (right lobe) in the right parasternal line. On percussion, its dullness could be separated definitely from that of the spleen.

Of course, the attacks of pain, one of which was followed by jaundice, were very suggestive of gallstones. On the other hand the anomalous character of the tumor mass attached to the left lobe of the liver, the fact that he had had syphilis, and that he had had, without obvious cause, two convulsive seizures, made me suspect that possibly the mass on the liver was syphilitic in character. He was ordered thirty grains of iodide of potassium three times a day.

On the 16th of April I saw him again, after he had had for a week or more a very severe attack of nausea and vomiting. The mass referred to was very evident. A feature of interest was the development, about a month after beginning to take iodide of potassium, of an acute parotiditis on the right side, probably secondary to the abdominal disease, such as Stephen Paget has so well described.

I saw the patient again on September 25th. He had had no attacks, no jaundice, no pains, and had not had a convulsion for five months. He has taken the iodide at intervals. Lately he has had an ulcer on one tendo Achillis, which was very troublesome, but is now healing. He has gained in weight, has been able to attend to his work, and looks very well. The tumor mass which was so perceptible in the left hypochondriac region has almost disappeared. The edge of the left lobe of the liver can be distinctly felt.

May. 3d.—He reports that he has kept well all through the winter. He has had no attacks of abdominal pain, no convulsions. He has gained in weight; looks well. The condition of the liver is practically negative. Nothing definite to be felt in the left lobe, only slight irregularity as it descends in deep inspiration.

IV. CANCER OF THE LIVER.—With the exception of the fibromyoma of the uterus, cancer of the liver may constitute the largest tumor met with in the abdomen. In extreme cases, as in the photographs I here show you, the entire cavity is occupied by the enormously enlarged liver. The diagnosis is not difficult, particularly in the secondary

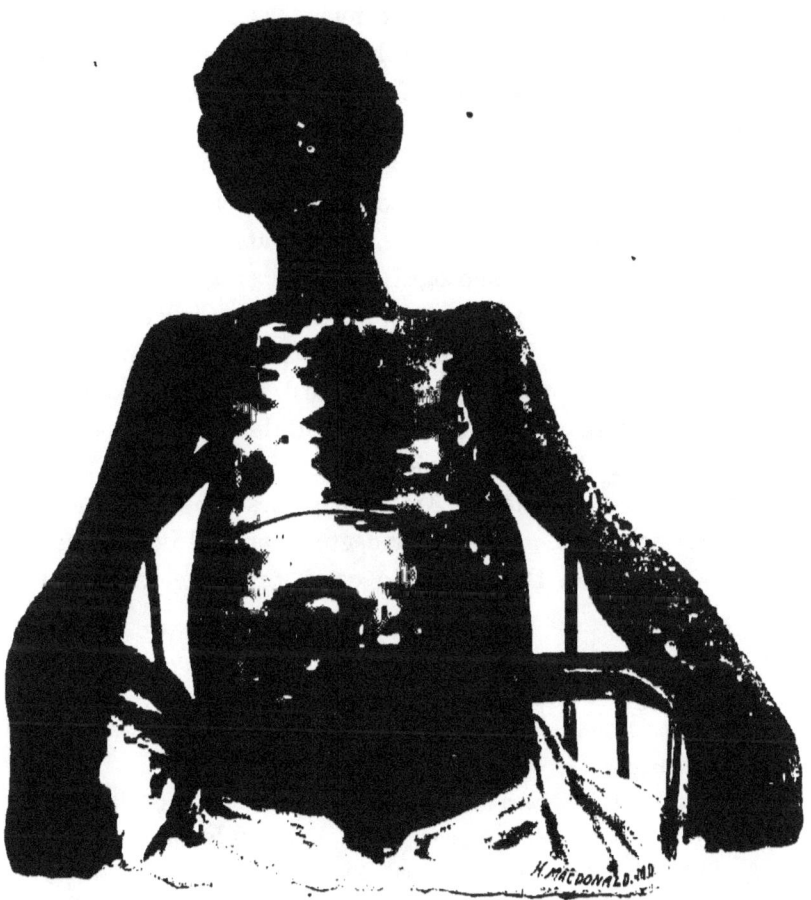

FIG. 28.—Cancer of the liver, showing the enormous increase in the area of dullness. The shaded areas show the situation of the visible tumor masses.

cancer with great enlargement of the organ. Cases of primary cancer, and especially the peculiar form of "cancer with cirrhosis," may be extremely difficult to recognize. A very large proportion of all cases are secondary, and characterized by a very rapid growth, profound cachexia, and often jaundice. The new growth may be so diffusely scattered throughout the organ that the enlargement is uniform and the surface is smooth; but more commonly there are large outgrowths on the surface or at the edge of the liver, which form prominent tumors of the greatest value in diagnosis. Not infrequently, indeed, they project beyond the surface of the liver far enough to be seen through the thin abdominal walls, as in this photograph from a case in the hospital last year (Fig. 28). In a patient with cancer of the liver (secondary to disease of the cæcum) who recently died on the surgical side, these masses were of great prominence, as this photograph shows (Fig. 29). These nodules are known in the older literature as Farre's tubercles of the liver. They vary in size from a walnut to an orange; they are usually firm and hard, the edges rounded, and the centers cupped—a sort of umbilication caused by the fibroid and degenerative changes going on in the central portion of the mass. Frequently these characters can be determined on palpation, and are of special importance in diagnosis. In the following cases tumor masses were present:

CASE XXXII. *Cancer of the Liver; Chills, Fever, and Sweats.* —Mrs. S., aged sixty-nine years, seen with Dr. Amanda Norris on April 30th. Family history good; husband died in 1879 of cancer of the stomach.

The patient has been a strong, healthy woman. During the past winter has not been in good health, has been losing in weight, and has had indigestion. She has kept about, and the condition was not thought to be serious until March 4th, when she had a severe chill with pain in the right side. There was no cough, no signs of any pleural or pulmonary trouble, nor had she any jaundice. The chills recurred very frequently, sometimes every day; she was in bed for three weeks. The fever rose, and she had heavy sweats. She had also marked gastric symptoms, and vomited almost every day; never brought up any blood, and never very large quantities. She got somewhat better, and for a couple

FIG. 29.—Showing the nodular masses of cancer of the liver.

of weeks was up and about and seemed improving, and she had no chills. During the past three weeks the chills and fever have been again present, and she has had vomiting and inability to take much food. The temperature goes up to 102° and 103°, and the chills are sometimes severe. The bowels have been constipated; she has never had any jaundice.

Present Condition.—Patient is not cachectic-looking; face has a grayish, rather anæmic aspect. Tongue is slightly furred. Examination of the thoracic organs is negative.

The abdomen is somewhat full; panniculus moderately thick; the upper zone of the abdomen a little prominent. On palpation, it is everywhere soft on the left side and below. Occupying the upper umbilical and all of the epigastric regions and extending to the right in an oblique line toward the anterior superior spine there is a solid resistant mass, which descends on inspiration. It is evidently a greatly enlarged liver. In the anterior axillary line the edge descends unusually low, and here, at a short distance from the anterior superior spine, there is a prominent nodular mass as large as the top of a lemon, firm, hard, attached to the liver. It gives one the impression of a secondary nodule. A second mass, not so large, can be felt just above the border of the liver in the nipple line, and a third at the edge of the liver, a little beyond the left upper sternal line. The upper limit of the liver dullness is greatly extended, particularly in the nipple and anterior axillary lines. The stomach does not appear to be dilated. No tumor mass to be felt on deep pressure in the left hypochondriac and umbilical regions; the spleen is not enlarged.

As usual, the chills in this latitude had been taken to indicate malaria, and she had been saturated with quinine for weeks without any influence. Chills with enlarged liver mean in the great majority of cases suppuration, either abscess or pylephlebitis. Here in a woman the onset with pains and the early chills suggest, even in the absence of jaundice, that the whole trouble may depend upon gallstones, and that the chills may be associated with suppurative cholangeitis.

Chills and fever may, however, occur in cancer of the liver, and in this case the emaciation, the enlargement of the organ, and particularly the nodular masses, suggest the presence of a neoplasm. The chills and fever may be associated with the rapid growth of cancer, but in the liver the suppuration may be in some of the large bile ducts, blocked with the neoplasm. Dr. Norris wrote that subsequently jaundice developed. The fever persisted, and before her death the emaciation was extreme.

CASE XXXIII. *Large Nodular Tumors at the Edge and Surface of the Liver.*—Mrs. S., aged about fifty years, consulted me January 24th, complaining of cough, loss of flesh, fever, and shortness of breath on exertion. There was slightly deficient expansion at the left apex, and a few râles in the suprascapular region. An examination of the sputum showed tubercle bacilli. I did not see the patient again until October 18th in consultation with Dr. Aaronsohn. She had had pleurisy on the left side, with some effusion, which had almost completely disappeared. She had become progressively weaker ; had had some loss of appetite, but no marked gastric symptoms. On examination of the abdomen, however, there was felt a remarkable ridge-like tumor extending just below the level of the navel, with a very hard, averted, and irregular edge, above which was a sort of shallow groove. The abdomen was much relaxed and the intestines lay between the abdominal wall and this ridge-like mass. At first I thought it possibly might be the omentum curled up and indurated, but on more careful palpation it was evident that the indurated, irregular edge was continuous with the liver. The extreme hardness and irregularity were, of course, very suggestive of cancer, in favor of which also were the enlargement and the pain on pressure.

I saw this patient again in consultation on the 30th and 31st, and the two weeks which had elapsed had made a very striking change in the condition of the liver. It was considerably below the level of the navel. The irregularity was very much more pronounced, and definite nodular masses could be felt both at the edge and on the surface. One of these, a little to the left of the middle line, was at least six centimetres in diameter, with a rounded edge and a depressed center. The condition was still a little peculiar and unusual from the fact that the abdominal walls were extremely relaxed and the intestines lay in front of the liver so that there was resonance as high as the costal margin. The growths in the liver were, from their local character, evidently secondary, and though the patient had profound anorexia, there was no evidence as to the seat of the primary disease. She died a few days after my last visit.

CASE XXXIV. *Enlargement of the Liver; Prominent Mass in the Upper Umbilical Region; Latent Cancer of the Stomach.*—Henry T., aged fifty-nine years, admitted October 4th, complaining of pain in the abdomen and back. Family history is good.

Has been a temperate man and has had no serious illnesses. Three months ago he says he was quite well. About eight weeks ago noticed that he had occasional pain in the abdomen, which for the past four weeks has been constant and of a dull aching character. He only stopped work three weeks ago ; has lost, he says, thirty pounds in weight in two months. His appetite is poor ; has never had any vomiting ; has no nausea. Food makes no difference in the pain. Two days ago his feet began to swell.

Patient is a tall man, much emaciated. The skin has everywhere a sallow tint, and the conjunctivæ are slightly tinged. Tongue moist, covered with a white fur. Condition of thoracic organs is negative. Abdomen much distended in epigastric and hypochondriac regions, especially on the right side. In the upper part of the umbilical region there is a prominent mass which is to be seen readily, and which moves up and down with respiration. On palpation, it is felt to be separated by a distinct groove from the swelling in the right hypochondriac and epigastric regions. The surface is smooth, painless ; no nodules are to be felt, but on the lower margin which extends to the navel it is distinctly irregular. The percussion dullness does not correspond to the edge of the mass, but is fully a hand's breadth above it. The upper limit of dullness is at the fifth rib in the nipple line, and at the seventh in the midaxillary. The splenic dullness is not increased ; the edge is not palpable.

The urine was dark brownish-yellow and contained a faint trace of albumin.

There seemed no question at all that this was a liver enlarged by cancer. but at first the prominent mass in the umbilical region, which seemed separated from the upper part by a distinct groove, raised a slight doubt ; but the profound cachexia, the rapid growth, and the irregular, nodular edge seemed conclusive. The primary trouble was not evident. The examination of the rectum was negative. A test breakfast, withdrawn fifty minutes after, gave fifty cubic centimetres of a dirty reddish-brown fluid consisting of undigested food, and showed a great many blood-cells. Free hydrochloric acid was not present. On the 10th he had been suffering a great deal of pain, and following three injections of a sixth of a grain of morphine at 9 A. M., 3 P. M., and 10 P. M., he became profoundly comatose, and died at 2 A. M. on the 11th.

The autopsy showed the primary carcinoma to be in the stomach, at the greater curvature, just eight centimetres from the cardiac orifice. The liver was enormously enlarged and weighed five kilogrammes and a half. The prominent tumor in the upper umbilical region felt during life corresponded to a new growth in the left lobe of the liver, which formed a projecting knob ten by ten centimetres in extent. The entire organ was occupied with small and large secondary nodules, very little liver substance remaining. The bile ducts were not affected. There were secondary nodules of cancer in the pancreas.

The following case is of great interest from the local character of the tumor masses, which in the epigastric region were so prominent, soft, and fluctuating that the condition of abscess of the liver was suspected. It illustrates, too, the importance of obtaining a thorough history.

CASE XXXV. *Sarcoma of the Liver; Two Prominent Tumor Masses in the Epigastric Region; Diagnosis of Abscess; Exploratory Laparotomy.*—E. K., aged nineteen years, seen September 6, 1892, with Dr. McGill. Condition on visit was as follows : The most extreme grade of emaciation, particularly of the face. The skin was bathed in perspiration ; pulse, 104, of fair volume and good tension ; respirations quiet ; no fever.

On exposing the abdomen, the upper zone is distinctly full, and two tumor masses are visible in the middle line, the smaller and less prominent just below the ensiform, and the other, a larger hemispherical mass, bulges the thin, tense skin between the ensiform cartilage and the navel. Both rose and fell with the respiratory movements. No glandular enlargements were visible.

96   THE DIAGNOSIS OF ABDOMINAL TUMORS.

On palpation, the superficial tumor masses were not tender, nor were there any spots of special sensitiveness anywhere over the liver. The lower and larger mass was soft and appeared to be distinctly fluctuating. The upper tumor was not quite so soft, and fluctuation could not be obtained between the two. The apparent fluctuation was also recognized by Dr. Tiffany, who had seen the patient some days before. A distinct ridge, like the edge of the liver, could be felt two fingers' breadth above the navel and extended to the right, passing at the anterior axillary line beneath the costal margin, at which point there was a somewhat indistinct irregularity.

The liver dullness began on the midsternum opposite the sixth costal cartilage, and extended within two fingers' breadth of the navel. In the midaxillary line it was at the eighth rib and the dullness was not increased at the right infrascapular region.

The condition of the heart and lungs was negative. The digestion was good and he had been taking plenty of nourishment. Lately he had had occasional attacks of diarrhœa.

The history of the case was not very satisfactory. He had been a fairly healthy lad, but had some indefinite illness this summer, and had gone out to Colorado with a friend. He was there on a ranch, and seemed to be fairly well until about six weeks ago, though he had apparently been losing in weight. He became much worse after a long ride, and about three weeks ago his father was summoned and immediately went to Colorado and brought him home. Since his return the chief symptoms have been progressive weakness and loss of flesh. The liver was found to be enlarged, and the tumor masses above referred to have within the past ten days become very prominent. There have been no definite chills, though he has occasional chilly feelings. The temperature has on no occasion been elevated and not infrequently been subnormal. He has had heavy sweats, particularly during sleep. No history could be obtained of any attack like dysentery, though he has had looseness of the bowels from time to time.

The first glance at the emaciated form of the patient at once suggested a new growth, but the age, the quick onset, and more particularly the examination of the superficial tumor masses and their rapid increase in size, seemed to favor the existence of abscess.

Suppurating hydatid tumor could not be definitely excluded, though the rapid course was against this idea ; also the profound emaciation which, though rare, is occasionally present, as in the case of an Italian who came under my observation in Montreal.* I suggested the propriety of aspiration or of an exploratory incision, and this the next day Dr. Tiffany proceeded to do. I then learned for the first time that in May, 1891, more than eighteen months ago, the lad had had disease of one testis, which had been removed, and Dr. McGill states that on section it seemed to be in a sloughing condition. He had, however, bruised himself on his bicycle. This fact was of very special importance in the history of the case, as it seemed most likely that the liver condition was associated with the disease of the testis, and from the length of time which had elapsed since the removal of the organ it rather favored the idea that the condition was neoplasm. I must say, however, that the physical examination of the two tumor masses in the epigastrium led us all to expect fluid, and I should unhesitatingly have put in an aspirating needle with the expectation of withdrawing either pus or a clear fluid.

Dr. Tiffany made an incision four inches in length over the lower tumor and exposed a large hemispherical swelling in the left lobe of the liver. There were no adhesions ; the superficial substance had a natural reddish-brown color, and puncture with the hypodermic needle withdrew nothing but blood. Dr. Tiffany inserted his fingers and examined the upper mass, which was a second soft enlargement, and on the under surface of the liver there were several others, leaving no question that there was a multiple new growth in the organ. The patient was extremely weak after the operation, but rallied for a few days.

In this case, as in one or two others which I can call to mind, I have been led astray by the deceptive, semi-fluctuating character of liver tumors.

Primary new growths in the liver in young men are, of course, extremely rare, and, taking all the circumstances of the case into account, it is more rational to suppose that

---

* *American Journal of the Medical Sciences*, October, 1882.

the lad had a new growth in the testis, which was bruised by the bicycle, and it was this in an inflamed condition which Dr. McGill removed in May, 1891.

The presence of tumor masses on the liver is, then, one of the most distinctive features of cancer of the organ, more particularly of the secondary form, which constitutes so large a proportion of all cases. The primary lesion is to be looked for in the stomach, intestines, urogenital organs, or the breast. The new growths are scattered diffusely with large nodular masses on the surface or at the edge. The rounded margin and cup-shaped depression are pathognomonic of these secondary cancerous nodules. The irregular syphilitic liver could alone be confounded with it, but in this condition there is rarely progressive enlargement of the organ, and the general features of the case are those of cirrhosis of the liver.

Tumor masses, as a rule, are absent in the primary cancer of the organ and in the form known as cancer with cirrhosis, in both of which conditions the organ may be of normal size, or even somewhat reduced. Lastly, large, rapidly growing encephaloid or sarcomatous growths may, as in Case XXXV, produce prominent tumors evident beneath the skin in the epigastric region, and which may apparently fluctuate, due either to the very soft nature of the neoplasm, or in some instances to hæmorrhage.

LECTURE IV.

TUMORS OF THE GALL BLADDER.*

THE gall bladder may be dilated or its walls infiltrated with a new growth. In a large proportion of all cases these conditions are associated with gallstones. Of six cases in which the gall bladder presented either tumor or enlargement, three were due to gallstones, one to compression of the common duct by malignant disease, and in two the walls were infiltrated with cancer.

(*a*) DILATED GALL BLADDER.—The organ may form a small, firm, rounded projection beneath the edge of the liver, a pyriform tumor of varying size and freely movable, in exceptional cases a very large tumor reaching to the pelvis, or, indeed, as in a case reported by Tait, a huge cyst occupying the greater part of the abdominal cavity. The usual causes of dilatation are blocking the ducts with calculi and compression of them by new growths. The greatest dilatation is associated with obstruction of the cystic duct. Permanent blocking of the common duct does not necessarily lead to very great distention of the gall bladder. The contents of the dilated organ may be a clear mucoid fluid when the obstruction is in the cystic duct and very prolonged; bile most frequently when the obstruction is in the common duct; pus or a puriform bile-stained fluid when suppuration has occurred, and an albuminous or bloody fluid in cancer of the walls. I will first

* Delivered December 6th.

read to you the cases in which the gall bladder formed a prominent tumor. Two of these were associated with gallstones and one with obstruction of the common duct by cancer. The diagnosis, which seemed perfectly clear, was confirmed in one case by operation and in another by autopsy. But before I do so let me call your attention to two monographs, the most important contributions to the literature of cholelithiasis which have been made for some years. Professor Naunyn's work, *Klinik der Cholelithiasis*, deals particularly with ætiology and symptomatology; while that of Professor Riedel, *Erfahrungen über die Gallensteinkrankheit, mit und ohne Icterus*, is of very special value to workers in clinical medicine, and illustrates in an interesting way the close interdependence of medicine and surgery. His careful study of an extensive series of cases upon which he had operated enlarges in certain directions our knowledge of the symptomatology of gallstones, more particularly of the cases without jaundice.

CASE XXXVI. *Gall Bladder forming a Visible Tumor; Operation; Recovery.*—Elizabeth D., Lonaconing, Alleghany County, Md., aged sixty-two years, seen with Dr. Kelly. She had been married thirty-eight years; had six children; three miscarriages; labors non-instrumental, but tedious.

Family history good. Has always been well until the present trouble. In December, 1891, she fell against the curbstone and struck the right side. At the time she felt very little pain, but remarked that she had had a great "shaking up." Some time in January, she does not know exactly when, she noticed a lump in the abdomen which changed in position as she moved in bed. It has not, she says, got any larger since January. Five weeks ago she one night became jaundiced. She is positive that the skin was clear one day and the next morning pretty deeply jaundiced. The stools were clay-colored, and remain so. Since then the bowels have been at times loose, sometimes three or four in a day. The urine became high-colored; the skin has been itchy, and the pulse

TUMORS OF THE GALL BLADDER. 101

rate very slow. She has lost somewhat in weight; she has had *no chills, no fever, and no sweating.*

Present Condition.—Large, well-nourished woman; the skin of an orange-green color; tongue clean; pulse, 32; temperature normal. The abdomen is enlarged and flabby. On the right side, midway between the costal margin and Poupart's ligament, there is to be seen a hemispherical prominence, which moves up and down with respiration. On palpation the abdomen is everywhere soft, not painful. The liver can be felt in the epigastric, right umbilical, and right hypochondriac regions, firm and resistant, and in the middle line the margin can be felt about two fingers' breadth above the navel. In the right parasternal line a distinct notch can be felt, the separation probably between the right and left lobes. In the nipple line the edge reaches to the level of the navel and in the anterior axillary line nearly to the level of the anterior superior spine. The prominent mass which is seen with such distinctness can be felt a hand's breadth below the liver margin; it is smooth, rounded, resistant, and as the fingers are pushed beneath it there is the impression of a globular body. It can be freely moved from side to side, and changes in position as she turns to the left. The dotted line in the diagram indicates the position when she rolled over on the left side; the fundus of the gall bladder then almost reaches the middle line.

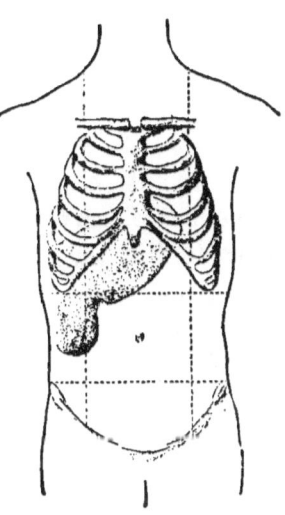

FIG. 30.—Showing the position of the gall bladder in Case XXXVI.

The surface of the liver is smooth. There are no nodules. On deep inspiration the spleen can not be felt; there are no glandular enlargements. There is a systolic murmur at the apex of the heart, but the sounds are clear at the base. The stools are clay-colored; the urine contains much bile pigment.

November 7th.—This morning Dr. Kelly made an incision fifteen centimetres long over the tumor. On opening the peritoneal cavity the liver looked of a dark greenish-brown color. Projecting beneath the edge of the right lobe for a distance of about five centimetres was the rounded end of a dilated gall bladder. The liver substance above it was considerably atrophied. There were no adhesions. The chief bulk of the dilatation was beneath the liver, and the dilatation was much greater than indicated by the portion which could be felt projecting beyond the edge. One hundred and fifty cubic centimetres of turbid, grayish pus were removed with the aspirator. Calculi could be felt in the cystic duct and at the first portion of the common duct. After aspiration the gall bladder was carefully stitched to the external wound and incised and a large gall stone removed weighing thirty-eight grammes. The stone occupied the cystic duct and projected into the common duct.

The points of interest in connection with this case are, in the first place, the easy diagnosis of dilated gall bladder on account of the position and character of the tumor. It seemed most likely, too, that the dilatation and the jaundice were due to gallstones, though she had never had attacks of biliary colic. In all probability the cystic duct was blocked by the large stone as early as December, when she fell against the curbstone; she is positive that the lump on the right side has been present ever since that time. The sudden onset of the jaundice five weeks ago was connected doubtless either with the moving of the stone into the common duct or the extension of inflammation from the cystic duct to it; most probably the former.

A second point of great interest in the case is the existence of an empyema of the gall bladder without chills or fever. In all probability the suppurative process was confined to the gall bladder and had not extended to the general bile passages, associated with which, so far as I know, there is invariably fever of a septic type.

The patient did very well. Much bile-stained material escaped from the wound, the jaundice became distinctly lighter, and bile appeared in the fæces. By the 14th of November the skin was much less yellow, the urine lighter in color, and the itching of the skin had entirely ceased. She improved rapidly, sat up by November 21st, and was discharged on December 14, 1892.

CASE XXVII. *Attacks of Gallstone Colic; Tumor of Gall Bladder.*—Miss S., aged about forty-eight years, seen with Dr. Ames, June 16, 1893, complaining of swelling and pain in the abdomen. Patient has been delicate from childhood, and has been for years a chronic invalid.

When about ten years of age she had a severe illness and for some time could take no nourishment without severe abdominal pain. Dr. Buckler thought that it was possibly ulceration of the stomach. When twenty years of age she had a similar illness— evidently protracted, painful dyspepsia. When thirty-five years old she had a severe attack of liver colic, in which she changed color and was slightly jaundiced. She has had since that time many attacks of pain, particularly at night, after a very trying or exhausting day. At forty-three the menopause began, following which she had a nervous breakdown and went to Italy, and for a couple of years she had a great deal of intestinal trouble. After returning, on September 6th of last year, she had a very severe attack of colic and was extremely weak. These attacks of colic have occurred throughout the winter and the last one she had was only a few weeks ago.

Present Condition. — Small-framed woman, looks ill, very anæmic and sallow ; no definite jaundice ; no special emaciation. Pulse 92; tension a little increased.

Abdomen flat, natural-looking. Palpation is everywhere negative until one reaches the liver region. Here inside the nipple line there is a definite rounded projection, the outlines of which can be readily determined, particularly below and to the left. It projects as a somewhat conical mass and is rounded at the right border. It is a little painful on deep pressure. The fingers can not be inserted definitely beneath it, but on either side the edge of the liver is distinct; as she draws a deep breath the fingers seem to pass over a prominence into a depression on the surface of the liver just at the level of the costal margin. During the attacks the projection, as she calls it, forms a prominent tumor which can be seen beneath the skin and is then exquisitely sensitive. The liver dullness extends to the upper border of the sixth rib. The right kidney is distinctly palpable and descends below the edge of the liver, from which it can be readily separated.

At a subsequent examination the tumor was not nearly so large, and she insists that it is extremely variable in size. Whenever the colic is severe the tumor becomes very much more prominent, a point confirmed repeatedly by Dr. Ames. She is excessively anæmic, and though anxious for an operation, it was thought best to postpone it until the condition of her blood was more satisfactory.

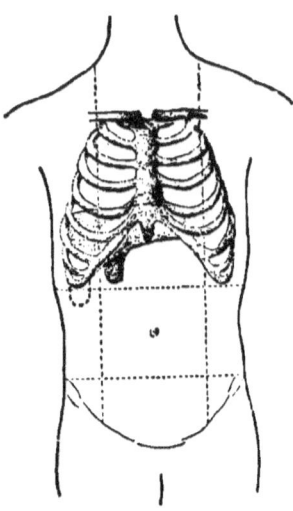

FIG. 31.—Showing the position of the gall bladder and outline of the right kidney.

Additional Note. — The tumor mass in the epigastrium developed considerably and became very much firmer, reaching almost to the level of the navel. The patient herself said that it had undergone a change, and she had a good deal of pain at intervals. She was very urgent that an operation should be performed, and, as her general condition seemed fairly good, Dr. Tiffany, on January 9th, made an exploratory incision. The gall bladder projected between two and three inches below the margin of the liver. A coil of the small intestine, and also the transverse colon, were firmly adherent to it, and over its surface there was a definite vascular membrane. There were numerous adhesions between the upper surface of the liver and the diaphragm. The gall bladder felt firm, and at its fundus there was a nodular mass. When it was opened a bile stained mucus exuded and the whole organ appeared filled with a semisolid, friable, grayish-yellow mass. Fragments removed with a curette, particularly the scrapings of the wall, had a grayish-white appearance, and looked like a new growth. On microscopical examination they consisted of large, irregular cells, many of which were in a state of fatty degeneration. After doing well for three or four days the patient began to complain very much of pain in the lower part of the abdomen. There was no swelling, and until the evening of the 14th there was no fever. On the 15th the pulse was more rapid, and she seemed very much weaker. She was worried about not having had an action of the bowels, and was a good deal exhausted in an attempt to get them relieved. About one o'clock she became feebler, gradually sank into a condition of unconsciousness, and died about 5.30 in the afternoon.

Autopsy, Jan. 16.—The wound of the skin and the incision in the gall bladder had both united well. After turning back the skin flaps, the intestines were found to cover the gall bladder everywhere, except at a small area just below the liver margin in the nipple line. The portion of the right lobe of the liver adjacent to it was considerably elongated, and consisted of a grayish-white atrophic tissue. The stomach was distended and the duodenum was pushed forward and much dilated with gas. It was opened *in situ*, and its posterior wall found to be in close contact with the gall bladder. The bile papilla, which was in the transverse portion of the duodenum, was not occluded. After dissecting off the duodenum the tumor beneath it was seen to be the greatly distended gall bladder, somewhat larger than a closed fist. It was deeply placed, lifted the head of the pancreas and the duodenum, as already mentioned, and was in contact with the right side of the vertebral column. It was readily and freely lifted from its bed, and then was seen to project about eight centimetres below the liver margin. The transverse colon was adherent to it, and just at the point of their union there was a nodule of new growth of the size of an English walnut. On incising the gall bladder, a small quantity of turbid bile exuded, but the entire viscus was occupied by a globular new growth, which was everywhere free, except at the upper wall, where it was densely adherent and had grown into the substance of the right lobe. It did not extend to the neck of the gall bladder, and the orifice of the cystic duct was free.

On section, the mass consisted of a fresh, grayish neoplasm, on its surface much bile-stained. There were six or eight small, black irregular gallstones. There were no secondary nodules in the liver.

The common duct was free. The stomach and intestines looked normal. There was a small cyst of the left parovarium, which was united to the rectum by old, firm adhesions.

CASE XXXVIII. *Enlarged Gall Bladder; Jaundice; Cancer of the Head of the Pancreas.*—Mr. M., aged fifty-one years, seen April 26, 1893, with Dr. W. W. Johnston. The patient is a large man, very active in business, and with an excellent family and personal history. He had a slight attack of jaundice during the civil war. Has not been a heavy drinker. Was well and strong until toward the end of last year. He had a mental shock and worry in October which distressed and disturbed him a good deal, and he had dyspepsia on one or two occasions before Christmas. Early in January he had a very severe attack with vomiting, and then began to lose in weight and had uneasy sensations in the epigastric region, but no sharp, acute pain. About the middle of January he noticed that he was yellow in color, and the jaundice, increasing in intensity, has been permanent. The stools have been clay-colored; the urine very much bile-tinged. He has had no itching, nor has the pulse ever been slow. He has lost progressively in weight, thinks as much as forty or fifty pounds, and has become very weak, though he has kept up and about, and until two weeks ago has attended to his business.

Present Condition.—Intense olive-green jaundice; moderate emaciation; pulse, 82; fair volume; moderate tension; vessel not sclerosed.

Abdomen prominent, and percussion and palpation demonstrate the existence of moderate ascites, which, Dr. Johnston says, has not materially changed for several weeks. On palpation, no pain, nothing abnormal to be felt until toward the right costal margin, below which the liver extends for about an inch and a half in the parasternal line. In the anterior axillary line, about two inches in front of the cartilage of the tenth rib, there is a firm, rounded, nodular body the size of the top of a lemon, which is attached to the liver and projects definitely beyond its edge. It feels like a distended gall bladder, but it is unusually hard, firm, and inelastic, and it is not movable from side to side. On deep inspiration, the surface of the liver above it can be felt to be distinctly depressed. In the parasternal line the edge of the liver is irregular, and there appear to be one or two nodules. In the middle line, on deep inspiration, the surface of the left lobe also appears rough and irregular. The spleen is not palpable. The stomach does not appear to be dilated. Examination of the thoracic organs is negative.

Patient died May 7, 1893. The autopsy showed cancer of the pancreas, with secondary nodules in the liver. The gall bladder was greatly distended and projected beyond the edge of the liver, and formed the tumor which had been so plainly to be felt during life, measured about six inches in length and about three inches in diameter, and was full of a light-greenish fluid. The walls were not indurated.

In the diagnosis of the tumor caused by dilatation of the gall bladder there are details to which I may here refer. The patient should be recumbent, in a perfectly easy posture, with the abdominal walls as much relaxed as possible. Sometimes, as in Case XXXVI, a prominent tumor is at once visible, descending with inspiration, or there may be a swelling of considerable size in the right half of the abdomen. More frequently, however, inspection is negative, and the facts must be elicited by careful palpation. The facility with which this procedure can be carried out depends upon the degree of rigidity of the abdominal walls, and a thorough examination may be impossible without anæsthetizing the patient. Bimanual palpation is the most satisfactory. Sitting by the side of the patient, the left hand beneath the lower ribs, with the right upon the abdomen, a little below the costal margin in the nipple line, gentle palpation with the pads of the fingers is first made during quiet breathing. The patient is then asked to draw a deep breath, and gentle but firm palpation is repeated, the fingers of the right hand following the receding abdominal walls. The anterior edge of a normal liver can in this way be readily felt, and any marked projection of the gall bladder detected. On the whole, I think you will find it more satisfactory to use the fingers of the right hand for palpation, but it is possible also to use the thumb of the left hand in the method described by Glénard, his *procédé du pouce*. The left hand grasps the right flank with the fingers behind. With the thumb, which is then free, the edge and surface of the anterior part of the right lobe of the liver can be readily felt, as the organ descends during inspiration. The facility with which this procedure can be carried out depends somewhat upon the length and mobility of the thumb.

Situation and General Characters.—The position varies with the size of the tumor and the existence of enlargement of the liver. Moderately distended in a liver of normal size, the gall bladder may be felt projecting beneath the costal margin opposite the end of the tenth costal cartilage. It is superficial, appearing to lie immediately beneath the abdominal wall. The long axis may be parallel with the nipple line. More frequently, however, the direction is somewhat to the left, as indicated by Fig. 31. The tumor is usually to the right of the parasternal line, but it may be directly in or even to the right of the nipple line, while in other instances it may be chiefly to the left of the parasternal line. The fingers may be placed directly beneath it, and the sensation given is that of a smooth, rounded body, larger at the lower end than above —that is, pear-shaped. While the outlines below are usually readily defined, toward the liver they are obscure, and no definite edge can be felt above the tumor. This is a point of importance in the differentiation of floating kidney. Sometimes the tumor appears to be turned forward on its axis, like a gourd, and a groove may be felt separating it from the liver. As a rule, palpation is not accompanied with much pain. The sensation conveyed to the finger is usually that of a tense, firm, elastic body. This is not always the case, for an enlarged gall bladder may be extremely flabby and soft, and is then difficult to feel. On the other hand, it is to be remembered that in long-standing cases of cholelithiasis there may be complete calcification of the walls of the gall bladder, forming a tumor of stony consistence. The size of the tumor projecting beyond the liver margin is some measure of the degree of distention of the gall bladder, particularly when the dilatation is due to plugging of the cystic duct. When the common duct is obstructed there may be great dilatation of the gall bladder and ducts with only a slight tumor projecting beyond the costal margin. I have reported a case, operated upon by the late Dr. Agnew, in which the fundus of the gall bladder projected only 2·5 centimetres, but on lifting up the liver it was seen that the distention was chiefly beneath the margin, and eighteen ounces of bile were removed by aspiration. While these statements hold for moderate dilatation of the gall bladder, you must remember that there are instances on record in which the tumor is exceptionally large, extending to the pelvis, occupying the entire right side, or even filling the abdominal cavity like a large ovarian cyst.

You will have noticed in the reading of the report of Case XXXVI that the tumor was extremely mobile and that the patient herself noticed its variability, and, projecting as it did so plainly, the alterations in position could be seen. The mobility during respiration is also well marked and it may be seen to descend with inspiration. On palpation the tumor may be moved freely from side to side. On the deepest inspiration, however, it can not be grasped and held in position as is possible with an extremely mobile kidney.

Nature of the Contents.—In a doubtful case of tumor projecting below the right costal border aspiration may be practiced, using a fine needle and exercising caution that the bowel does not lie between the tumor and the abdominal wall. The contents of a dilated gall bladder are either clear mucus, which is most common in prolonged obstruction of the cystic duct; bile, when the common duct is blocked, though when occluded for a prolonged period the entire bile passages, including the gall bladder, may be filled with a thin mucus. Pus is met with frequently, usually general symptoms indicating that suppuration has occurred; and, lastly, blood may be present in cases of neoplasm of the gall bladder. In acute phlegmonous inflammation due to calculus, the gall bladder may contain a dirty, brownish-red, ill-smelling fluid.

From growths at the pylorus and in the colon, which may occupy a similar position, the gall-bladder tumor is, as a rule, readily distinguished, both by the differences in the symptoms, and particularly by the systematic local examination, using also the inflation of the stomach and intestine. To two conditions I would, however, call your attention. In stout persons and when the abdominal walls are unusually tense, movable kidney on the right side may be mistaken for an enlarged gall bladder. Only, however, when the kidney is very movable does it descend so low and so far to the left that this mistake could occur. It does sometimes, however, emerge beneath the liver margin as a rounded tumor in a most deceptive manner. With the patient recumbent and the kidney in its natural position, no tumor is evident; but on change of position (turning to left) or on deep inspiration it then appears. A movable kidney on the deepest inspiration, with the fingers placed above it, can be held down and prevented from returning during the expiratory movement. A gall-bladder tumor rises and falls with the expiratory movements, and can not be held down during expiration. Again, above the rounded surface of the kidney, the sharp margin of the liver may be felt with great distinctness, whereas in gall-bladder tumor the upper limit is not to be defined, and there is no sharp edge above it.

An interesting anatomical condition of the liver which you must learn to recognize has been referred to—particularly by Professor Riedel—namely, a tongue-shaped process of the anterior margin. A knowledge of its existence may save you from error. I show you here the outline as given by him in one of his cases (Fig. 32). He believes that this extension is seen particularly in women whose livers have suffered from the effects of lacing, but it is directly caused by traction, the gradually distending gall bladder elongating the anterior margin. In twelve of the cases upon which Riedel operated this tongue-like process was present; in nine instances the gall bladder was palpable either at the median or under margin of the process. In Case XXXVIII I believe this process is present. It is not always, however, associated with dilated gall bladder, and I have seen very curious elongations of the anterior margin of the right lobe in perfectly normal livers, and in several instances of the posterior margin of the left lobe. It is important to recognize the existence of this process as it may form a very definite mass in the right flank. A very interesting instance of it was referred to me by Dr. Weir Mitchell two years ago. An extremely nervous woman, aged about fifty-six years, had had for several years symptoms of neurasthenia, pains in the abdomen, and ill-defined manifestations, for which she had sought re-

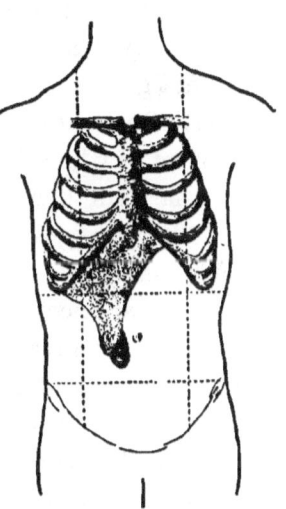

FIG. 32.—The tongue-shaped extension of the anterior margin of the right lobe, with the gall bladder projecting below it. (Riedel.)

lief in many quarters. In his note to me Dr. Mitchell stated that there was a tumor of a doubtful nature in the right flank. I saw the patient with her relative Dr. Tilghman. The only interest in the case is in the examination of the abdomen. Just beneath the right costal margin, extending toward the anterior superior spine, was an elongated mass with very ill-defined borders. The fingers could not be inserted beneath it, nor was there a definite edge palpable. It was tender on pressure, not movable in a lateral direction, but with the fingers deep in the flank behind it could be tilted forward. It did not seem to be continuous with the liver, the dullness of which terminated just below the costal margin. It descended somewhat with the respiratory movements. The right kidney was not palpable. Altogether I was puzzled by the condition, and could not give a positive opinion as to the nature of the trouble. She had a good deal of abdominal pain and distress, most of which I thought was associated with nervous dyspepsia. As there had been some uterine trouble I suggested that the pelvic organs should be examined. Subsequently, she came under the care of my colleague, Dr. Kelly, who at once discovered the tumor in the abdomen and, with the advantage pertaining to surgery, he very quickly determined by laparotomy the nature of the tumor —namely, an elongation from the right lobe of the liver. He writes: "There was no visceral trouble, excepting the enormously elongated thinned-out lobe of the liver, which extended down on the right side at least four inches below the normal position, and seemed to be an elongation of the anterior margin. The gall bladder was not enlarged." *

(b) CASES WITH ILL-DEFINED NODULAR TUMOR AT LIVER EDGE, SUPPOSED TO BE GALL BLADDER.—While the

---

* Those interested in the subject may consult with advantage Hertz's recent work, *Abnormitäten in der Lage und Form der Bauchorgane bei dem erwachsenen Weibe, eine Folge des Schnüren und Hängebauches.*

presence of a well-defined tumor is of the utmost importance in the diagnosis of gall-bladder disease, there are cases in which we have to be content with less positive evidence. A special value of Professor Riedel's work in relation to the diagnosis of gallstones lies in a somewhat startling revision of accepted data regarding the cardinal symptoms of this disease. Thus, ten of the fifty cases upon which he operated had never had colic; only fourteen presented a definite tumor, and a majority had never had jaundice.

The following cases are of interest from the fact that a small nodular tumor was felt in both, not very clearly defined; in one the history of severe and protracted attacks of colic seemed clearly to indicate the presence of gallstones, while in the other case the condition was more doubtful.

CASE XXXIX. *Severe Attacks of Colic for Five Years; Nodular Tumor at Edge of Liver; Operation with Removal of Three Hundred Gallstones.*—August W., aged forty-two years, seen April 28, 1893. Sent by Dr. Salzer.

Patient is a large-framed, well-built man; has always enjoyed excellent health, with the exception of malaria twelve or thirteen years ago. Has always had a very good appetite and has been a heavy eater. He has not been a dyspeptic, but has been troubled at times with constipation, but until the onset of his present illness he always regarded himself as a very healthy man.

He comes complaining of attacks of severe and protracted pain in the abdomen, which began five years ago. While walking in the garden one evening he had severe colicky pains, like cramps, which lasted throughout the greater part of the night; he was not jaundiced and the attack was regarded as one of simple colic. Six months afterward he had a second and more severe attack, which came on suddenly in the same way. Subsequently the attacks became more frequent, and he had at least six or seven in the second year, and they have gradually increased until he has been rarely a month or six weeks without pain. He has only once had vomiting with the pains, and never brought up any blood; has never had diarrhœa, most frequently has been constipated. The pain begins as a rule in the upper abdomen, radiates to the back, and sometimes is very diffuse throughout the back and sides. Its duration varies extremely; thus, the day before yesterday he had a severe attack; yesterday he was free from pain. In a bad attack he is quite incapacitated, and can not straighten himself to walk. He never has had bloody urine after the attacks,and never has passed gravel. The early part of this winter the colic was very severe and he lost considerably in weight. I could get no history of jaundice from the patient or from his wife, but Dr. Salzer informs me that at least twice icterus followed the attacks of severe pain.

Present Condition.—Face is flushed; color good; venules of cheeks a little dilated; conjunctivæ not stained.

Abdomen is full; panniculus well preserved; on palpation everywhere soft. In nipple line, three fingers' breadth below costal margin, he winces on pressure, and here is to be felt an irregularity at the edge of the liver. In middle line the edge is indefinite; in parasternal line the liver can be felt about two fingers' breadth below margin, and there is an ill-defined, rounded, somewhat nodular mass in this situation, which moves with respiration. It is sensitive on deep pressure. Liver dullness begins at the seventh rib and extends to the costal margin in parasternal, and a finger's breadth below in nipple line.

The edge of the spleen is not palpable, but dullness extends from the eighth to the eleventh ribs in midaxillary line.

The stomach is not dilated; thoracic organs negative.

The diagnosis of gallstones was made and he was advised to have an operation.

May 4th.—Dr. Halsted opened the abdomen and explored the gall bladder. It was found that the irregular mass at the edge of the liver was due to a marked projection of the anterior border. It was not to the left of the notch of the gall bladder and did not form a definite tongue-like extension, but was rather an irregular projection of the border. There were numerous adhesions between the under surface of the liver and the colon. The gall bladder did not look enlarged. It was laid open and found to contain a clear mucoid fluid and about three hundred gallstones of various sizes, chiefly very small, but one at the orifice of the cystic duct was the size of a small cherry. The patient reacted well from the operation, had no fever, and made a satisfactory recovery.

In the following case, led astray by a nodular prominence at the edge of the liver, I thought the condition was possibly gallstones in the common duct; but unfortunately the exploratory laparotomy did not give us any definite information.

CASE XL. *Enlarged Liver; Nodular Tumor at Margin; thought to be the Fundus of Gall Bladder.*—Henry L., aged thirty-three years, traveler for a spirit house, was admitted November 1, 1892, complaining of jaundice.

Father died at sixty years of paralysis; mother and seven brothers and sisters living and well.

Patient has always been well and strong. The only serious disease was diphtheria at his fifteenth year. Patient has been a pretty steady drinker, chiefly of beer; has had gonorrhœa, and was under treatment many years ago for syphilis, the constitutional symptoms of which were very slight. He has never had hœmorrhoids.

He has not been feeling very well for a year or more, and has been depressed in spirits owing to domestic troubles. He has been at work until a month ago. The present illness began two months ago, when he noticed that he was gradually getting yellow. There were no pains at the outset; no colic; nor had he nausea or vomiting. The appetite was good; the bowels were regular, and he did not feel badly enough to stop work for more than a month after the jaundice appeared. He had at times a sense of weight and dragging in the liver region, but never any pain. He has lost gradually about twenty pounds in weight within the past few months.

Present Condition.—Patient is a well-built, spare man; not specially emaciated; skin and conjunctivæ of a tolerably intense yellow color. The temperature is normal; pulse 64, full, regular; tension a little increased and the vessel wall slightly thickened. Thorax is barrel-shaped; resonance everywhere clear; no adventitious sounds. The heart beat is in the normal position; the sounds are clear, and the second, at the aortic cartilage, is decidedly accentuated.

Abdomen is enlarged, particularly in the upper zone, and is prominent in the right hypochondriac region. On palpation the whole of the right hypochondriac and epigastric, and part of the umbilical regions are occupied by a firm, resistant mass, corresponding to an enlarged liver. The lower border extends to within about three centimetres of the umbilicus and passes under the right costal margin at the junction of the eighth and ninth costal cartilages. The flatness begins at the sixth rib in the nipple line. The surface is, as a rule, smooth, though toward the navel slight irregularities are felt. The edge is not very well defined. A little within the nipple line, at about five centimetres to the right of the navel, there is felt on deep palpation a little projection, rounded, somewhat nodular, and which appears to be attached to the liver border. It is not movable, and is in the position and extremely suggestive of the tip of the gall bladder projecting beneath the liver margin. The edge of the spleen was just palpable on deep inspiration. The splenic flatness began in the eighth interspace. The stomach was not dilated and the gastric juice contained free hydrochloric acid.

The urine was bile-stained, clear, acid, 1·025, contained a trace of albumin and a few hyaline and granular casts.

The patient remained in hospital three weeks. The jaundice varied considerably in intensity, and the complexion got at times very much clearer. The stools were clay-colored, and at no time were of such tint as to indicate that at any rate much bile passed into the intestine.

During his stay in the hospital the patient gained a couple of pounds in weight, his appetite was good, and he was always able to be up and about.

The nature of the trouble did not seem at all clear. The patient's habits, the length of time the liver had been enlarged, the size of the organ, without ascites, favored the view that he had a form of hypertrophic cirrhosis, to which even the intensity of the jaundice was not opposed. The question of syphilis was also discussed. The stools, however, had rather the character of a definitely obstructive jaundice, and at times he was intensely yellow. There seemed a possibility that the common duct was obstructed, and, though he had not the intermittent fever and chills so common in the impaction of a gallstone in this part, yet the jaundice

TUMORS OF THE GALL BLADDER. 117

was variable, and the nodular mass at the edge of the liver was suggestive, to say the least, of enlargement of the gall bladder. He wished for an operation to determine the nature of the trouble, and agreed to return for an exploratory laparotomy.

January 2, 1893.—Patient came back to-day in practically the same condition; but he has gained a couple of pounds in weight. There is no change in the liver, and the nodular enlargement can still be felt. The skin is deeply jaundiced, the stools clay-colored, and the urine very dark.

6th.—This morning Dr. Halsted did an exploratory laparotomy. An incision was made about three inches below the costal margin, just above the border of the liver and following its curve. When the peritonæum was opened there appeared to view, covering the entire surface of the liver, a smooth structure, covered by peritonæum, looking like omental tissue, containing vessels and fatty tissue. It was adherent to, but could be moved like a skin upon the surface of the organ. The nature of it was doubtful. There were numerous adhesions between the edge of the liver and the transverse colon, which had to be separated, and the lower surface of the liver was united by adhesions to the adjacent parts. In the separation of the adhesions there was a good deal of bleeding, and the vessels had to be tied. The gall bladder was not found, and it was impossible under the circumstances to make a satisfactory dissection of the gastro-hepatic omentum. Nothing abnormal was felt about the head of the pancreas, and no stone could be felt in the duct. The edge of the liver itself was irregular, and at a little distance from the margin there was a distinct indentation or groove. The nodular mass which we felt so repeatedly was, in all probability, a projecting portion of this ledge-like edge. The membranous fold already mentioned was loosely adherent to the surface of the liver, and, when lifted up, there were numerous bleeding points which had to be touched with the Paquelin cautery. The surface was somewhat irregular, roughened, of an intensely bluish, almost plum color, and looked like an organ in a state of hypertrophic cirrhosis.

It did not seem possible to be able to determine precisely the nature of the remarkable fold of membrane covering the liver. It covered the left lobe and extended up as far as could be felt. It seemed more like a large properitoneal membrane covering the liver. It was outside of the liver capsule, which was not itself thickened.

I heard from this patient last on November 6, 1893. His jaundice had all disappeared, and the stools and urine are natural. He had improved a good deal, but the dropsy had been very much worse, and he had been tapped twelve times.

(c) CANCER OF THE GALL BLADDER.—New growths of the gall bladder, which are not very uncommon, have of late attracted much attention, particularly in their relation to gallstones. The diagnosis is in some cases easy, in others extremely difficult, and if the patient has had attacks of gallstone colic, and presents a rounded tumor mass below the edge of the liver, the condition is very naturally regarded as simple dilatation of the gall bladder. The following cases, which have been under observation, illustrate certain points in the diagnosis:

CASE XLI. *Persistent Jaundice with Emaciation and Ascites; Nodular Tumor at Edge of Right Lobe.*—Magdalen H., aged fifty-two years, admitted to Ward G on October 18, 1892, complaining of swelling of the abdomen and legs.

Her father died of tuberculosis. No history of cancerous disease of the family.

The patient has always been very healthy, was married at twenty-two; had one child. She has been troubled for many years with constipation. She has never had attacks of colic.

The present illness, dating from about the middle of June, began with vomiting, after which she became yellow and had itching of the skin. The jaundice has never entirely disappeared. The legs became swollen about the end of August, and the abdomen six weeks ago. There has been pain in the back, so that she always has to lie on the side; otherwise she has not had much distress. The stools have been yellow. She has had but little vomiting. There has been progressive loss of weight, and she has become very weak.

Present Condition.—Patient is much emaciated, and has an intense olive-green jaundice. There is general anasarca. The abdomen is extremely distended, and the lower zone of the thorax is expanded. Without going into details foreign to the main point, it may be said that she had all the signs of obstructive jaundice, and an ascites which required frequent tapping. The immediate interest of the case was in the condition of the liver. After tapping, the liver was distinctly palpable, and in the parasternal line the rounded edge could be felt about two finger breadths from the costal margin. Passing toward the flank, in the anterior axillary line, a prominent nodular mass was reached, and here the liver margin was nearly seven centimetres below the costal margin. The mass felt, about the size of a walnut, was prominent, not umbilicated. No other masses could be felt, but the edge of the liver in the parasternal line was somewhat irregular.

Remarks.—This illustrates a group of cases of obstructive jaundice the precise cause of which is often difficult to determine. The persistent icterus and the loss of weight suggest a new growth, but whether in the stomach, the pancreas, or the liver itself is almost impossible to say. A test breakfast shows free hydrochloric acid, and she has not had much vomiting since admission to hospital. The stools are grayish yellow, not fatty and not suggestive of pancreatic disease. The nodular body at the right border is the main objective point in the local examination, and the question discussed between Dr. Thayer and myself before the class has been whether this is a secondary nodular growth or the projecting end of a firm, hard, cancerous gall bladder. To my touch it rather resembles the former, feeling as though the finger could be passed all around the lvier tissue at its base. Supposing it to be secondary cancer of the liver, the organ is not nearly so large as is common in this condition in the space of five or six months. On the other hand, in primary cancer of the gall passages the liver is often not much enlarged, and the jaundice, as in this case, is intense from the outset. A point in favor of this view is the absence of evident signs of disease of the stomach, pancreas, or intestines.

Patient died November 16, 1892. The above comments were written before the patient's death. The autopsy showed a primary carcinoma of the gall bladder, the end of which was the nodular body which we had been able to feel so definitely on palpation. The walls of the organ were greatly thickened, and it contained nearly one hundred small gallstones. There was great induration and thickening about the common bile duct, the head of the pancreas, and in the gastro-hepatic omentum. The common duct passed through this indurated tissue and was almost occluded. The liver weighed only fifteen hundred grammes and presented numerous medium-sized cancerous nodules throughout its substance.

CASE XLII. *Cancer of the Gall Bladder; Jaundice; Progressive Emaciation.*—E. S., aged fifty-four years, admitted to Ward G on January 25, 1893, complaining of pain in the abdomen and soreness in the back. There is nothing of any moment in the family history. She has been married; has had six children; four miscarriages. She has never had uterine trouble ; no serious illness until the present attack.

More than a year ago she had pains in the back, sometimes quite severe, and accompanied with high-colored urine. After several of these attacks she had passed small calculi in the urine. She has had none of these attacks and has not passed a stone for about a year. She has been failing in health for the past few months, has had indigestion, belching, and occasional attacks of vomiting, and has lost a good deal in weight. About five weeks ago she noticed change of color in the skin and that she was getting yellow, and for about the same time she has had a dull aching pain on the right side of the abdomen. The urine has been high-colored, and the stools, which formerly were very dark, were light-gray in color.

The patient is a medium-sized woman; face thin, but the body and limbs still well nourished. There is moderate jaundice.

Abdomen full; panniculus well retained. On palpation, it is soft, nowhere painful except at a point about five centimetres below the costal margin in the nipple line. Here there is a firm mass which extends to the left to within six centimetres of the umbilicus, and at this border the fingers can be placed directly beneath it. Below it reaches to the transverse navel line, and is here rounded, and the fingers can not be placed so well beneath it as to the left. To the right the margins are not very clearly defined, but it extends nearly to the tip of the tenth rib. Above, it can not be separated from the liver margin. It feels like a rounded mass larger than a lemon, is extremely resistant, hard, and, though it has the situation of the gall bladder, it scarcely conveys the impression of the rounded, pear-shaped outline of that organ. The right kidney can not be felt. The liver dullness is not present in the midsternal line, just three centimetres and a half in the parasternal and five centimetres in the nipple line. The mass above described, though directly continuous with the liver, presents a flat tympany on percussion. Deep pressure from behind in the right flank presses the tumor mass forward.

The spleen is not enlarged, stomach not dilated, and the pelvis is clear. The urine is very dark in color, 1·016, pale; bile pigments present; no sugar; a few granular casts. The stools are clay-colored and very offensive. Repeated examinations showed no essential change in the condition of the tumor mass. The jaundice became very much more intense, though the general symptoms were somewhat ameliorated. She took her food better, and had much less pain.

The case was regarded as tumor of the gall bladder associated with gallstones, and probably malignant disease. The patient's condition was so satisfactory that it was thought advisable to have an exploratory operation to determine if anything could be done.

February 8th.—This morning Dr. Halsted made an exploratory operation. The mass above described was in the situation already referred to—between the transverse colon and the under surface of the liver, to which it was firmly attached. The adhesions to the colon were so tight that it was not thought advisable to attempt to separate them. The tumor mass was firm, solid, and grayish white in color, passed beneath the surface of the liver, and occupied the position of the gall bladder. The liver itself was not enlarged, but the edge could readily be felt about six centimetres above the lower border of the tumor mass.

The jaundice persisted: she got progressively emaciated; the wound healed. Her friends took her home on March 2d, where she subsequently died.

Carcinoma of the gall bladder is not very easy to recognize, but there are certain suggestive features in suspected cases. The disease is most common in women—two thirds of the cases collected by Musser. In seven eighths of the cases the cancer has been associated with gallstones, so that a history of colic or of previous jaundice should be sought for. Rapid emaciation, with or without jaundice, and the development of a cachexia within three or four months, speak for cancer; simple hydrops vesicæ may persist for months without impairment of the general health. Chills and fever are, as a rule, against neoplasm. So long as the disease is confined to the gall bladder jaundice is not present, but when it extends to the common duct the icterus is intense and persistent. Ascites may be caused by the propagation of the disease to the peritonæum by pressure of secondary masses on the vena portæ, by extension to the gastro-hepatic omentum, as in Case XLI, and occasionally is due to thrombosis of the portal veins.

The local features are variable and uncertain. In the cases I have narrated the walls of the gall bladder were infiltrated, but the cancer may be at the outlet, causing obstruction with great dilatation and a tumor resembling in all respects that produced by any other occlusion of the cystic duct. When the fundus is involved the tumor is harder, more resistant, not so movable as in simple hydrops, and the growth may be very rapid. The liver is not usually much enlarged, even when secondary nodules are present. Aspiration of the tumor gives most important indications. A clear mucoid fluid favors gallstones; turbid, albuminous contents suggest neoplasm, as does also blood or a blood-stained fluid. Fragments of the new growth may be found in the material aspirated. Pure bile is rather in favor of gallstones, and indicates that the cystic duct is not involved.

But taking all the circumstances, general and local,

into consideration, you may not be able to reach a conclusion, in which case remember that the hazard of an exploratory operation is slight, and that by far the most frequent cause of tumor in the region of the gall bladder is cholelithiasis.

LECTURE V.

TUMORS OF THE INTESTINE, OMENTUM, AND PANCREAS;
MISCELLANEOUS TUMORS.

1. *Tumors of the Intestine.*—Cancer, the common cause of tumor, occurs most frequently (apart from the rectum) in the cæcum and the sigmoid, hepatic, and splenic flexures of the colon. Not one of the three cases which have been under observation presented a typical group of symptoms, but singly and together they illustrate many interesting features of the disease. In the first place, the affection may be latent, revealed at autopsy alone, or the early and indeed the chief symptoms may be due to the secondary tumors. The first case illustrates very well the latency of the disease. Without intestinal symptoms, for some months after he came under observation the sole objective feature was the progressive enlargement of the liver. Only six weeks before his death, after he had become greatly emaciated, we discovered a tumor in the right iliac region, and subsequently he had hæmorrhage from the bowel.

CASE XLIII. *Cancer of the Cæcum and Colon; Latent Course; Enormous Secondary Enlargement of Liver.*—John R., aged twenty-nine years, admitted February 6, 1892 ; under observation until September 25th. The family history is good. The patient was healthy and strong until four years ago, when he had severe malaria. For eight months prior to his admission he had not been very well, and had had irregular pains in the abdomen. During the past eight months he has been pale, has felt weak, has not been able to work, and on several occasions he has been slightly jaundiced. His appetite has been good; bowels regular; has had no diarrhœa. For several months past he has noticed that the upper part of the abdomen was swollen and tender to the touch. On admission, the patient was pale, but looked well nourished; no fever; pulse 86. The lower thoracic zone is much expanded, particularly on the right side. The epigastric and hypochondriac regions bulge in a very prominent manner, and there is a rounded mass, nine centimetres in transverse extent, which extends from under the ribs on the right side. In the median line the edge is clearly defined, and reaches to within four centimetres of the navel. To the left it extends far over beyond the parasternal line, and to the right deep into the lumbar region. The percussion over this large mass is flat and continuous with the liver dullness, which begins in the median line at the base of the xiphoid, in the nipple line at the sixth cartilage, and in the axilla at the seventh. Although there was no fever and no definite history of any intestinal trouble, the patient's age and good condition seemed against the diagnosis of cancer of the liver. Accordingly an aspirator needle was thrust in at the prominent part, but only blood obtained. Under observation the liver evidently increased in size, and there seemed to be no question that it was a new growth. The question then arose as to the primary seat of the disease. The stomach symptoms were insignificant, he had no vomiting, the appetite was good, and a test breakfast was readily disposed of. Subsequently, in August, very careful examination of the abdomen revealed a hard mass low down in the flank. It was usually ill-defined, but on several occasions Dr. Thayer thought that it was quite distinct. There was no diarrhœa; no special change in the fæces, which were always well formed. He remained under observation outside the hospital during the summer. The liver tumor did not increase very much in size. He became progressively weaker and very much emaciated. Two weeks before death he passed two large stools containing clots of blood. He became extremely emaciated before his death.

Autopsy.—The liver weighed seven thousand two hundred grammes; the right lobe was much disfigured, and presented numerous nodular tumors with elevated margins and depressed centers. A distinct groove marked off the anterior margin of the right lobe from the rest of the organ. On section, secondary cancerous nodules were found scattered through the entire organ. The primary growth was found to be at the head of the cæcum and the beginning of the colon, which presented an extensive fungating mass, softened and necrotic on the surface. The mesocolon was thickened and the glands much involved. Microscopically the tumor proved to be a cylindrical celled epithelioma. Extensive secondary nodules were scattered through the lungs. .

In the following, by far the most interesting and instructive case of the series, intestinal symptoms were absent throughout, and the presence of a solid, firm mass deep in the right side led us to think at first that there was a renal tumor. I give you the notes just as I dictated them from day to day, as they illustrate the erroneous diagnosis, and the gradual development of features which led to its revision.

CASE XLIV. *Tumor in Right Flank; Suspected to be a Renal Sarcoma; Subsequent Development of Dilatation of the Stomach and Signs of Tumor in the Bowel.*—Thomas B., aged thirty-two years, admitted August 3, 1892, complaining of pains in the abdomen. There was nothing of special moment in his family history. Present illness began about eighteen months ago with griping pains in the abdomen, attacks of which occurred from time to time and were attributed to indiscretions in diet. On the voyage to this country, ten months ago, he was very seasick, and had a great deal of pain in the abdomen; and then for the first time he noticed a hardness or lump on the right side. The bowels were, as a rule, constipated. He has never passed blood in the stools or in the urine. For some months he has been gradually losing in weight, and has been getting pale and weak. When in good condition he weighs one hundred and seventy-seven pounds; he now weighs one hundred and twelve pounds.

Present Condition.—Patient is a tall, well-built man, pale, scarcely cachectic. The tongue is moist, slightly coated; pulse 104, of fair volume. The examination of the thoracic organs is negative. The abdomen is symmetrical, but looks a little fuller in the right flank. The right hypochondriac and lumbar regions are filled with a firm, somewhat irregularly rounded mass, which on bimanual palpation can be readily moved up and down. On making firm pressure with the left hand in the right renal region the mass becomes apparent beneath the skin just to the right of the navel. To the left it extends almost to the middle line; the lower border is three centimetres below the transverse navel line. Both the lower and the left borders are rounded, but toward the right, just beneath the tip of the eleventh rib, there is a distinct nodule to be felt. In its anterior part it can be separated distinctly from the liver both by palpation and percussion, but in the anterior axillary line the tumor passes beneath the ribs, and the dullness is here continuous with that of the liver. On percussion there is resonance over the tumor mass to the left of the nipple line. The general situation of the tumor is indicated in the annexed chart. There are no glandular enlargements. The spleen is not palpable and the dullness is almost obliterated.

The patient has no sweats, no cough. There is a leucocytosis, the white corpuscles numbering over twenty thousand to the cubic millimetre; hæmoglobin, forty-one per cent.; red blood-corpuscles slightly over four million to the cubic millimetre. The urine (many examinations) has usually been clear, acid; specific gravity, 1·012; at first no albumin, but subsequently slight traces. No sugar; microscopically, a few blood-cells, but as a rule, even after centrifugalizing, neither casts nor blood-cells were found. The sediment obtained was also examined for tubercle bacilli, with negative result. The temperature was at times a little above normal, and on the evening of the 20th of August he had a chill, the temperature rising to 103·5°. On the 6th of September a medium-sized aspirator

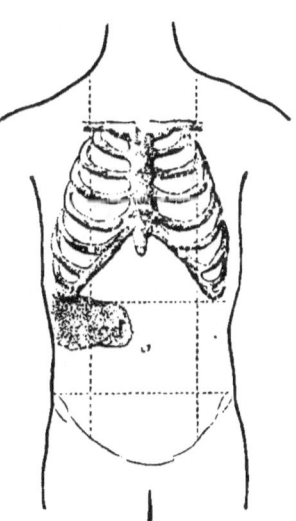

FIG. 33.—Outline of the tumor mass in Case XLIV.

needle was thrust upward and backward beneath the twelfth rib, and a little blood-stained fluid removed, which contained round cells twice the size of leucocytes. The needle seemed imbedded in the firm substance. The patient objected to an exploratory operation.

The situation and shape of the tumor, the mobility, and the readiness with which it could be pushed forward by pressure from behind pointed to a renal origin. The nature of the growth was less certain. If he is correct in dating the first symptoms as far back as eighteen months the tumor has not attained the size usually reached by new growths of the kidney within this time. There has also been no blood in the urine. The single chill does not speak specially against new growth. Can it be tuberculous nephritis ? The family history is good ; the urine is and has been clear; there is no involvement of the epididymus, and fever has not been a marked feature. If the tumor were due to saccular dilatation following occlusion by tuberculous or calculous disease, certainly the aspirator would have withdrawn purulent fluid, and there would have been at some time pus in the urine. The rapid loss in weight points strongly in favor of new growth.

October 12th.—The patient was shown in clinic this morning. The emaciation has progressed; he has had for ten days much more fever, and a chill on the sixth in which the temperature rose to nearly 104°; on the seventh, eighth, and ninth it remained between 102° and 103°. There are no special changes in the urine. He has been at times constipated, but the stools show nothing peculiar. The tumor mass has not increased materially in size, though perhaps it reaches a little further toward the navel.

The patient says that he notices flatus bubbling in the vicinity of the tumor, and that it divides into, as he expresses it, two or three portions. The hand placed on the tumor experiences occasionally a feeling as if gas was escaping through it, and the left half is resonant; but this might be due to the presence of the colon over the mass.

This morning, with the students, the various probabilities of renal sarcoma, or tuberculosis, or calculous pyelitis were discussed. The state of the urine and the failure of aspiration to draw fluid seem opposed to the latter conditions. The chills and fever were not thought to be inconsistent with sarcoma.

The question was also discussed as to whether it really was a renal tumor, and whether it might not be associated with the liver or with the hepatic flexure of the colon. It did not seem possible, with the evidence at our disposal, to reach a definite diagnosis.

Since the above note of October 12th there have been several developments in this case.

15th.—For the past two days the patient has had a great deal of vomiting, often bringing up large quantities. Last night he vomited eight hundred cubic centimetres of brownish fluid containing half-digested food. The reaction was acid, the odor rancid, and tests for free hydrochloric acid were negative.

16th.—Ewald's test breakfast given this morning, and withdrawn fifty minutes later, gave nearly three hundred cubic centimetres of a slightly grayish, muddy fluid, containing comparatively little food matter. It was acid in reaction, odor rancid and sour, turned congo paper blue, and gave a very distinct rosy-red color with the phloroglucin-vanillin solution. Microscopically, there were fat crystals, undigested food stuffs, numerous bacilli, and yeast cells.

17th.—The vomiting has continued during the past twenty-four hours, and he does not look so well to-day. The tumor mass occupies the right hypochondriac region, extends into the right lumbar and umbilical regions, but not into the epigastric, reaching apparently to within about two centimetres of the navel. Below, it extends exactly eight centimetres from the costal margin in the nipple line. The greatest prominence is just below the point of the tenth rib. On first palpating it this morning there was at the lower margin a prominent rounded, ridge-like mass, firm and hard, which gradually disappeared, feeling as if it were a tubular, muscular structure in contraction. Again this morning gas was felt bubbling through the mass. Percussion over it gave flat tympany; slight change noted in rolling the patient over on the left side. At a second visit to-day there were noticed for the first time waves of peristalsis crossing the upper abdomen from left to right, and the

130   THE DIAGNOSIS OF ABDOMINAL TUMORS.

outlines of the stomach could be distinctly seen, the lower border reaching to the navel. At the time of the passing of the waves the walls of the stomach became firm, and bubbles of gas could be felt passing through the tumor.

20th.—The signs of dilatation of the stomach have been for the past few days unusually distinct. He has had vomiting of large

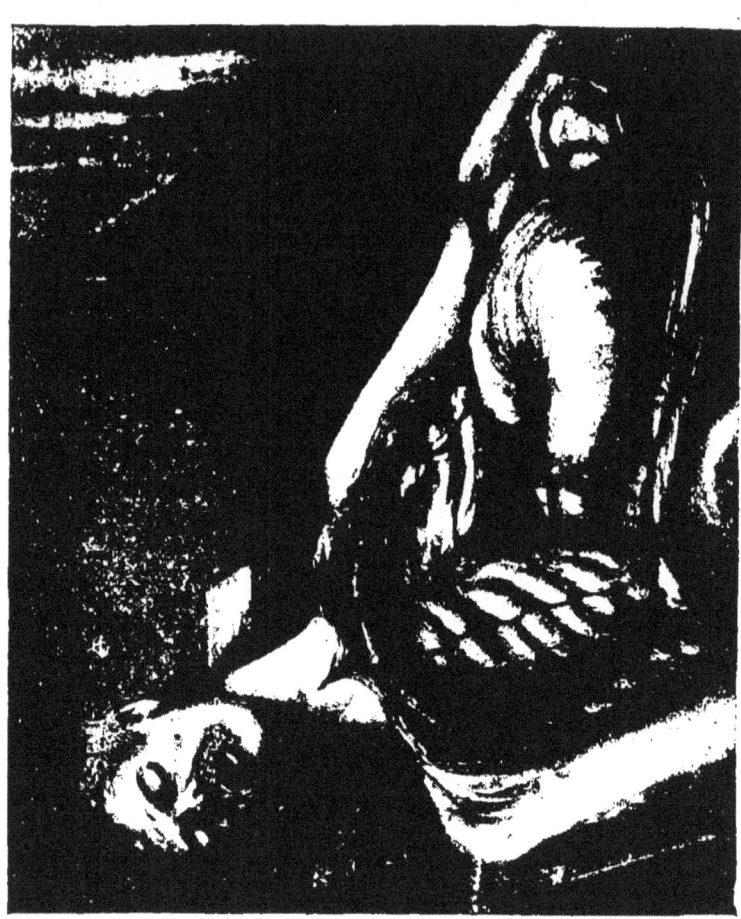

FIG. 34.—Cancer of the colon; dilatation of the duodenum and stomach in Case XLIV.

quantities of liquid. The amount of urinary secretion is very scanty. He sank gradually and died on the 20th.

Autopsy.—By Dr. Flexner. On opening the abdomen the stomach was seen to be greatly dilated, reaching considerably below the level of the navel (Fig. 35). In the right hypochondriac region a tumor mass was adherent in part to the anterior abdominal wall, just between the costal margin and the crest of the ilium. Just below the hepatic flexure of the colon there was a tumor the size of an orange completely encircling the bowel. It was seven centimetres in length and eight centimetres in circumference. The coats were uniformly infiltrated and the tissue looked infiltrated with colloid. The inner surface was ulcerated and the lumen of the bowel not here narrow. The cæcum and ascending colon were opened *in situ*. At the hepatic flexure the tumor mass was adherent to the right lobe of the liver and behind was attached to the kidney. The finger, introduced into the colon at this region, entered a number of pockets, one of which directly led into the liver substance. On its posterior surface and to the right the tumor was closely united to the curve of the duodenum, into which it had grown in such a way as to cause a distinct narrowing. The mucous membrane of the duodenum was ulcerated from the central part of the tumor, but was intact elsewhere. The stomach was greatly dilated; the mucous membrane smooth. The pylorus itself and first part of the duodenum were greatly stretched; the groove between them is well seen in the figure. The finger could be passed into the duodenum, but the narrowed lumen would not more than admit the tip of the little finger. The liver showed numerous secondary nodules of cancer. The mesenteric glands were enlarged and contained metastatic nodules.

The third case presented a very prominent movable tumor which, from its general characters and situation, seemed to be connected with the bowel, though the intestinal symptoms were here also quite in the background.

CASE XLV. *Tumor in Right Flank; Removal of Growth in Cæcum and Ascending Colon.*—Sylvester H., aged sixty years, admitted October 15th, complaining of a lump in the abdomen. He has been a very healthy man. For two years past has had slight pain after eating, with nausea and constipation.

His present illness began about three months ago with troublesome constipation, and three or four days would pass without a movement from the bowels, and the fæces would be hard and lumpy. He took medicine for it, since which time the bowels have been rather loose, the stools yellow and containing slime, but never blood. No tenesmus, no cramps. Five weeks ago he had an attack of vomiting, followed by a second attack a week later; brought up sour material; no blood. In the past three months he has lost a great deal in weight, and has become, he thinks, a little pale. He has been short of breath on exertion, and lately the feet have been swollen.

Present Condition.—Patient looks a little pale, but is fairly well nourished; no fever; tongue slightly furred. Abdomen is full, and about six centimetres to the right of the navel there is a projection beneath the skin, uninfluenced in position by the respiratory movements. On palpation this is felt to correspond to a rounded nodular mass, feeling of about the size of a cricket ball, situated midway between the navel and the anterior superior spine of the ilium. It feels very superficial; is not tender; is very hard; and one or two ridges can be felt upon it. There are no changes in its consistence. On deep inspiration it descends slightly. It is freely movable and can be pushed over as far as the navel. No gas is to be felt to bubble through it. On light percussion there is a flat tympany over it.

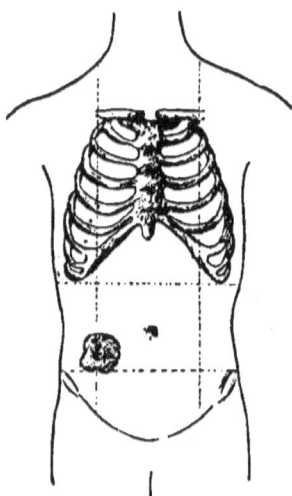

Fig. 35.—Situation of the tumor in Case XLV.

As the patient had had very few intestinal symptoms and had had dyspepsia for several years, with recently two attacks of vomiting, the attention was naturally directed, in the first place, to the condition of the stomach. Palpation was negative in the region of the pylorus. There was no *clapôtage*. The organ did not appear to be dilated. The upper limit of resonance was at the fifth rib in the nipple line, and the lower limit above the navel. On October 17th the tests for free hydrochloric acid were negative. On the 21st it seemed that the stomach tympany was somewhat more extensive than before. There were no peristaltic waves. On several occasions a test breakfast was given, and on October 26th the mucus withdrawn was blood-stained ; no free hydrochloric acid. The tumor could be readily separated from the liver and could be moved far down into the right iliac fossa. While the symptoms in this case pointed rather to disease of the stomach, the situation and general character of the tumor were those of an intestinal growth. The stomach, too, seemed somewhat relaxed, and the absence of free hydrochloric acid was suggestive.

On November 7th Dr. Halsted operated and found that the tumor occupied the cæcum and the commencement of the ascending colon. It was readily removed. The patient seemed to do very well, taking his nourishment and gaining in strength until the 13th, when, after an attack of nausea and coughing, the stitches gave way and about two feet and and a half of the small intestine protruded. He became very restless, gradually sank, and died the same day. The tumor was a cylindrical-celled epithelioma, involving the entire circumference of the gut, but excavated and not narrowing in any way the lumen of the gut. The autopsy showed an interesting feature—namely, the presence also of a cancer on the posterior wall of the stomach and of a second small tumor mass in the jejunum.

In the diagnosis of cancer of the intestine the following points may be taken into consideration. In comparison with the subjects of malignant disease of the stomach very many of the patients are young; thus you have noticed that Case XLIII was only twenty-nine years of age, and Case XLIV was only thirty-two. Intestinal features are present in a majority of cases, though they were by no means suggestive in the patients who have been under our observation. Griping, colicky pains are common, even without the signs of obstruction. With narrowing of the lumen of the gut very characteristic features occur—attacks of severe griping pain, abdominal distention, the presence of active, sometimes visible peristalsis in the distended coils of bowel, and, if the condition persists, vomiting and all the signs of intestinal obstruction.

In the case of Sylvester H., I called your attention repeatedly to the fact that the intestinal symptoms depended largely upon the state of the lumen of the bowel at the seat of the tumor. If fungous masses project and cause more or less narrowing, colicky pains and constipation are inevitable; but, on the other hand, as the tumor grows, if there is necrosis of its surface, with excavation, neither pain nor constipation may be present. Diarrhœa and the passage of much slime with the fæces are not infrequent symptoms. Hæmorrhage is also common. The blood is not often in large quantities; when the tumor is in the sigmoid flexure it may be bright and very little changed, but in growths about the cæcum it is often much altered before it appears in the stools. There are cases in which constant loss of small quantities of blood is a very special feature, and the patient becomes profoundly anæmic. Sloughy fragments of the tumor may sometimes be passed in the fæces.

A cachexia develops progressively but with very variable rapidity. It may, however, be well marked before any features have arisen suggestive of intestinal trouble. The loss in weight may, too, be slight, even after the tumor has persisted for many months. There is at present a patient in Ward C with a tumor in the right iliac region, which has persisted for nine or ten months, and upon the nature of which very many opinions have been expressed. She is well nourished, but profoundly anæmic. It did not seem possible from the symptoms, general or local, to make a definite diagnosis, but an exploratory operation showed an extensive new growth in the cæcum. Another patient, who had repeated small hæmorrhages, developed an extreme anæmia with retention of the general fatty panniculus. When extensive secondary growths develop, as in Case XLIII, the cachexia may be profound. The tumor in cancer of the intestines may be readily and easily discovered—indeed, evident on inspection, as in Case XLV. On the other hand, as in Case XLIII, it may not be until the terminal stage of the disease that the growth is found. A small tumor of the hepatic or splenic flexure of the colon may escape repeated examinations. Mobility is a special feature of growths in the large bowel. Large tumors, however, of the cæcum may be quite fixed. The most movable growths are those connected with the sigmoid flexure. Variability in size is also a marked character, and at one examination the mass may appear as large as the closed fist or even two fists, and the next day it appears not larger than a small apple. These variations are due largely to the presence of fæcal masses in the vicinity. Two very important features in the intestinal tumor may sometimes be detected on careful palpation—namely, the hardening during contraction of the hypertrophied wall in the vicinity of the growth, and the bubbling of gas through the tumor, which may be heard as well as felt. This latter feature drew our attention to the possibility of the tumor n Case XLIV being associated with the colon. The intestinal symptoms above referred to and a progressive cachexia are generally sufficient to warrant a diagnosis.

II. OMENTAL TUMORS.—In two cases a rolled, thickened omentum formed a definite tumor in the upper portion of the abdomen. I will not enter into full details, but just mention the cases in abstract.

CASE XLVI. *Pleuro-peritoneal Tuberculosis; Ridge-like Tumor in the Epigastric Region.*—A man, aged thirty years, admitted May 9, 1893, with ascites. In October he had had shortness of breath with swelling of the legs and abdomen. He improved gradually, but the ascites has persisted and on admission there were signs also of effusion in the right pleura. In the abdomen there was an ill-defined resistance, a hand's breadth in width, at the junction of the umbilical and epigastric regions; below it terminated in a well-defined, hard border, which could be very easily felt, and was indeed at first thought to be the edge of the liver. On percussion there was a flat tympany above the hard transverse ridge. There was in this region also a very well marked peritoneal friction rub. The history of the case, the involvement of pleura and peritonæum, and the existence of this transversely placed tumor mass in the upper abdominal zone led to the diagnosis of tuberculosis. Dr. Finney made an exploratory operation and drained the peritonæum. The omentum was rolled up, thickened, and attached to the transverse colon. The patient did well and was discharged from the hospital greatly improved.

CASE XLVII. *Chronic Proliferative Peritonitis; Thickened Omentum.*—In this patient the tumor was more interesting and unusual, though I must say no diagnosis was made. I refer to it especially because I have been talking to you so much in the ward class about chronic proliferative peritonitis in connection with the case of the little girl with a visibly pulsating liver. Without going into unnecessary details, the patient, aged about fifty-five years, was admitted with suppurative cellulitis of the left leg which had come on in connection with an ascites of some weeks' duration. His condition was very serious, and one for which we could not do very much. He gradually sank, and died ten days after admission. The abdomen was persistently distended, and we never arrived at a definite opinion as to the cause of the dropsy. The parietal peritonæum was adherent to the anterior surface of the colon and to the omentum for a distance of 3·5 centimetres. The omentum was represented by a thick fold, seven centimetres in vertical by nineteen centimetres in transverse extent, the upper part of which was converted into a white, shining, leathery-like structure, not, however, rolled or curled upon itself. Similar thickenings were present over the anterior surface of the colon. The intestines, particularly the loops of small bowel, were bound together by dense adhesions, separated with the greatest difficulty, and there were patches of thickening on the mesentery. There was a condition of chronic perisplenitis and perihepatitis. There was a thrombus in the portal vein.

As in the chronic tuberculous peritonitis, this simple proliferative form may pucker the omentum into a definite tumor, lying athwart the upper zone of the abdomen. Encapsulated exudate may also form tumor-like masses. More frequently, however, the recurring ascites simulates cirrhosis of the liver. In one way the proliferative peritonitis may produce a very extraordinary tumor, of which I have reported an example.* So great may be the thickening of the mesentery that the whole bowel is shortened, and the coils of intestine matted together may form a mass the size of a cocoanut firmly bound to the spine.

III. TUMORS OF THE PANCREAS.—Two cases of disease of this organ came before me for diagnosis—one a cancer, the other possibly a cyst.

The cancer case you will remember, as I demonstrated the specimens after a ward class early in the session. A reasonable probability may sometimes be reached in the diagnosis. The following may be mentioned as suggestive points: Rapid emaciation with early, intense, and persistent jaundice; dilatation of the gall bladder, fatty stools, glycosuria, salivation, and the presence of a tumor between the umbilicus and ensiform cartilage. Nausea and vomiting, though often present, are variable features. Beginning usually in the head of the pancreas, the growth early compresses the common duct and causes obstructive jaundice. A persistent, intense icterus may also result from compression of the duct in the gastro-hepatic omentum by infiltrated glands, secondary to cancer of the stomach, to invasion of the ducts themselves by cancer, and by stenosis of the common duct, rarely by gallstones without any complication. The emaciation is rapid, and stress is laid by some writers upon an excessive cachexia. There may be very little pain throughout the illness; in some cases, however, attacks of colic occur. A dilated soft gall bladder, while not in itself of special im-

---

* Tuberculous Peritonitis, *Johns Hopkins Hospital Reports*, vol. ii.

port, is often suggestive, taken in connection with other features. In the primary cancer of the bile passages, which also causes an early and intense icterus, the gall bladder, if enlarged, is more often hard and firm. Disturbance of the function of the organ may be manifest by (1) the presence of fat in the stools (even a definite stearrhœa), which is not a constant symptom, but of value when present; (2) by glycosuria, which is also not constant. Salivation is sometimes present, and Dr. Lainé, of Media, has called my attention to several cases in which this symptom was present, and in which the diagnosis of cancer was confirmed by autopsy.

The tumor in cancer of the pancreas is not always to be felt, and, as in Case XLVIII, there may be ascites, which renders it difficult. The mass may be between the navel and the margin of the right lobe of the liver, as in our case. It is deep seated, not mobile, variable with the degree of distention of the stomach and intestines. The involvement of the adjacent parts may give to it the characters of a deep-seated, dense, massive tumor. Bear in mind that in very thin-walled persons, particularly in women with enteroptosis, the pancreas can be felt with distinctness, but the conditions are very exceptional in which it could be mistaken for a tumor.

CASE XLVIII. *Intense Jaundice; Progressive Cachexia; Ascites; Tumor in Epigastric Region.*—E. V., aged thirty-four years, admitted to Ward C, October 24th, complaining of dyspepsia and jaundice. Family history is good.

With the exception of chills and fever fifteen years ago, he has been a very healthy man. For a year he does not think he has been in his usual health, feeling tired and out of sorts; but has had no nausea and no vomiting. For about two months he has been losing rapidly in weight, and has had uncomfortable sensations in the abdomen after eating. A month ago jaundice developed, and has gradually become very intense. He has had no severe pain.

TUMORS OF THE PANCREAS. 139

The upper part of the abdomen has been a good deal swollen. He came to hospital on account of the jaundice, weakness, and progressive emaciation. Patient presented all the characters of severe obstructive jaundice. He is much emaciated; skin everywhere of a deep yellow color; no fever; pulse, 76. The abdomen is distended, particularly in the upper zone. It is tympanitic in front and dull in the flanks, with well-marked movable dullness. No peristalsis seen in epigastric region. The prominent tympanitic zone extends as low as the umbilicus. On deep pressure in the right epigastric region, between the navel and the costal margin there is a hard nodular mass, difficult to outline, owing to the distention. There is enlargement of the lymph glands. The test breakfast showed the presence of free hydrochloric acid, and the stomach did not appear to be dilated. Examination of the heart and lungs was negative. On October 28th and on November 1st he had some stomach distress, for which lavage was practiced, and a quantity of dark, coffee-ground-looking material was washed out. On November 2d it was noted that the distention in this case was unusual; no coils were to be seen; no peristalsis. The tympany extended to the fifth rib on the left side, and a little above the costal margin on the right side. The stools were clay-colored, rather firm, and he constantly had to take purgative mineral waters. Microscopically there was sometimes a good deal of fat. There was never any sugar in the urine, which had the usual characters of this secretion in obstructive jaundice.

The intensity of the jaundice, the rapid emaciation, without enlargement of the liver or recognizable disease of the stomach, and the presence of a deep-seated tumor mass led to the suggestion of pancreatic disease. The tumor was difficult to feel satisfactorily owing to the very great distention of the epigastric region, and as the patient's physician had suggested that it was possibly due to gallstones, and as he was himself very anxious that something should be done, Dr. Halsted performed an exploratory operation. More fluid was found in the peritonæum than we had expected. The remarkable distention in the epigastrium was due to the floating up on the top of the fluid of the colon and small bowel. The gall bladder was found to be dilated; but the mass which had been felt was a deep-seated growth in the situation of the head of the pancreas. The patient was a good deal relieved by the operation, but no essential change took place, and he gradually sank, becoming very intensely emaciated, and died November 18th.

The post-mortem by Dr. Flexner showed the head and body of the pancreas to be the seat of a tumor mass. The growth had infiltrated the wall of the duodenum, and the posterior wall of the stomach was involved directly from the tumor in an area eight centimetres in extent. The gall bladder and ducts were much dilated with dark, thick bile.

To the case of possible cyst of the pancreas I shall only just refer, and show you the chart, as Dr. Halsted, in whose practice it occurred, will publish it with full details.

The man (Case XLIX), aged about thirty years, was admitted, April 14, 1893, with a greatly swollen abdomen, measuring over forty inches in circumference. His illness dated from January, 1890, when, without any fall or injury, he had for three days severe colic, not associated with vomiting or with jaundice. A month later he had a second attack, also lasting three days; in this one he vomited, and noticed for the first time swelling of the abdomen. Then the attacks recurred frequently, two or three a month, each time with nausea, vomiting, and colic, and the abdomen progressively enlarged until July, 1890, after which he had no further attacks of colic. The abdomen remained large, but his general condition was good and he was able to do light work. In July 1892, he fell out of a wagon, jumped up, got into it again, but immediately had a severe attack of colic, and had to go to bed in a hotel near by for two days. He had nausea, vomiting, and great pain. The swelling gradually disappeared, and in ten days

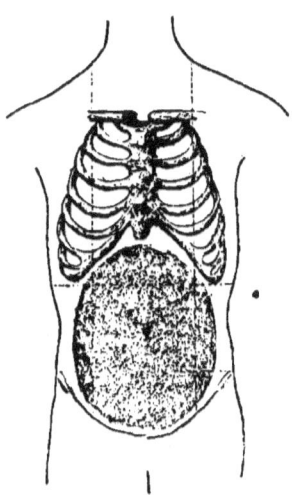

Fig. 36.—A cyst in the abdomen, probably of the pancreas (Case XLIX).

the girth of the abdomen decreased from forty-three to thirty-one inches. He had profuse diarrhœa, but he does not think there was pus or blood in the stools. He gained in weight rapidly, and went to work. He was married in October, and remained well until January of this year, when he felt his trousers were tight at the waist ; and during the past three months, without colic, vomiting, or jaundice, the abdomen has gradually increased in size.

A truly remarkable history! The man was unusually intelligent, and insisted that his statement about the complete disappearance of the tumor after the fall from a wagon was correct. When he came under observation the entire abdomen was distended, particularly in the upper zone ; the ensiform cartilage and the lower ribs were everted. The distance from the tip of the ensiform cartilage to the navel was 21·5 centimetres ; from the navel to the pubes, 16·5 centimetres. The wall was tense and elastic; fluctuation was readily obtained. The percussion outlines are given in the chart. There was resonance only in the epigastric angle, in the hypochondriac regions, and in the flanks. In the right inguinal region there was an elevated ridge. Dr. Halsted incised and drained the cyst, which was found adherent to the abdominal wall. Eighteen litres of a dark, coffee-colored fluid were removed, which was alkaline in reaction, contained granular *débris* and much altered blood, fresh red blood-corpuscles, and a few large cells deeply stained with blood pigment.

The patient had a tardy convalescence, but ultimately left the hospital in good condition. The diagnosis of pancreatic cyst in this case rests rather on general than special grounds. The gradual development with attacks of colic, the persistence without serious damage to the health, the disappearance once after a fall, the gradual reaccumulation —all point to a retention cyst. There were no features pointing to loss of the function of the pancreas—neither fatty stools nor glycosuria. Large pancreatic cysts may fill, as did this one, the entire abdomen ; in long-standing cases the contents consist, as a rule, of altered blood, and while an amylolytic ferment, which is not in any way distinctive, may exist, the trypsin, which would be definite and conclusive, is usually not present.

Many of the cases described as cysts of the pancreas are really instances of hæmorrhage into the lesser peritonæum. Let me refer you to Mr. Jordan Lloyd's suggestive and timely contribution.* From four of his conclusions you may gather the gist of the whole matter.

1. "That contusions of the upper part of the abdomen may be followed by the development of a tumor in the epigastric, umbilical, and left hypochondriac regions."

2. "That such tumors may be due to fluid accumulations in the lesser peritoneal cavity."

3. "That, when the contents of such tumors are found to have the property of rapidly converting starch into sugar, we may assume that the pancreas has been injured."

4. "That many such tumors have been regarded as true retention 'cysts of the pancreas,' and that this opinion has been formed upon insufficient evidence."

IV. MISCELLANEOUS TUMORS.—In this group I shall place seven cases, in two of which the diagnosis was doubtful or could not be definitely made.

(a) *A Cyst of the Abdomen of Unknown Origin (Mesenteric).*—The following case presents many remarkable features. For more than two years he has had recurring distention of the abdomen, which reaches such a size that he has to be tapped. With the exception of the period of onset, from October to December, 1891, during which he lost about fifty pounds in weight, he has remained in excellent condition, and is inconvenienced only by the bulk of the fluid as it accumulates. He has usually gone to work the day after the tapping. The physical signs are those of a cyst. The dullness is in the front of

---

* *British Medical Journal*, November 12, 1892.

the abdomen, with resonance in the flanks. He was tapped eight times in 1892, and five times up to date this year. The quantity removed has varied from one gallon to five gallons and a half. At the last two tappings the amount has been only a gallon and a gallon and a half. At first the fluid was dark and bloody, but since the second aspiration it has been a turbid, muddy-looking fluid, alkaline in reaction, containing leucocytes in a condition of disintegration, much granular and molecular *débris*, and very many cholesterin crystals. I regret that no chemical examination was made of the digestive properties of the sample of the fluid which was sent to me. Dr. Miller has reported recently that the patient is in the best of health, and the amount of fluid is gradually diminishing.

CASE L. *Cyst of Doubtful Origin.*—November 29, 1892. I saw to-day the following very unusual and remarkable case :

X. F., aged forty-four years. Referred to me by Dr. G. B. Miller, of Philadelphia. Patient is a large-framed, stout man, looking the picture of health.

Family history is good. His personal history is also excellent. He had the usual diseases of childhood; scarlet fever, but not a very severe attack. He has had two attacks of gonorrhœa; has never had syphilis. He has been married ten years; no children. He uses alcohol moderately; has never been a heavy drinker. At times during the past six or eight years he has had "gouty" pains about the joints. In October, 1891, he noticed that he was getting uncomfortably large in the abdomen, and for this he took three bottles of some "reduction remedy," and lost in weight everywhere except in the abdomen, which became progressively enlarged. He then consulted Dr. Loeling, who told him he had fluid in the abdomen. He kept at work, however, until December, feeling weak and having occasional attacks of nausea and vomiting. The distention of the abdomen became so extreme, in spite of active catharsis and diuretics, that, on December 26th, he was tapped and five gallons and a half of a dark, bloody fluid with drawn. An examination of the abdomen after removal of the fluid failed to reveal any hardness, tenderness, or tumor. The urine at this time was negative. He lost in weight from two hundred and fifty pounds to two hundred and two pounds. After the first tapping he gained in strength, and very quickly went to his business. Gradually, however, throughout January the fluid reaccumulated, and on February 14, 1892, he was again tapped, and three gallons of dark, bloody fluid removed. Without any aggravation of his general condition, and without any special interference with his business, the fluid continued to reaccumulate at intervals, and he was tapped on the following dates : March 25th, five gallons and a quarter; May 5th, five gallons and a half of dark serum; June 12th, one gallon; June 22d, dry tapping, no fluid was obtained; July 9th, twenty-eight pounds; August 12th, three gallons and a half of muddy, turbid fluid ; October 4th, twenty-six pounds of turbid, muddy fluid; November 22d, last tapping, three gallons removed. He has felt no inconvenience from the tappings, and has usually resumed work on the following day. The only trouble has been the gradual increase in the size of the abdomen, which causes shortness of breath on exertion and a feeling of tension. He has never at any time had swelling of the feet ; the bowels have been regular; the appetite has been lately very good, and, as a rule, with the exception of a short period this summer when he got pale and thin, he has been in very good health and has been able to attend to his work systematically.

Present Condition. —As stated, the patient is a large-framed, powerfully built man, looking the picture of health. The color is good ; the venules on the cheeks are somewhat marked; the tongue is clean; the pulse is quiet, 78 a minute; tension moderate; no sclerosis of the arteries.

The abdomen is moderately full, but not larger than is frequently seen in a man of his build. It is symmetrical, not specially prominent in any region. The pulsation of the abdominal aorta is not transmitted to the surface; respiratory movements of the abdomen are natural. There is no special enlargement of the superficial veins. The inguinal glands are not enlarged. On palpation it is everywhere soft and painless. No tumor masses or areas of specially increased resistance are to be felt. On deep palpation below the right costal margin, and during inspiration, the edge of the liver can be touched. Fluctuation can not be obtained.

On percussion the entire front of the abdomen is flat, and only on the deepest percussion in the region of the navel is there flat tympany. The dullness continues into the left flank, but there is a flat tympany high up beneath the tenth and eleventh ribs, and this is continuous in the left hypochondriac and left epigastric areas with the stomach tympany. In the right flank, between the costal margin and the ileum, there is resonance. On turning from side to side, resonance on the left side becomes more extensive, that on the right side not much changed. In the nipple line the liver dullness begins at the lower margin of the sixth rib, and there is no tympanitic note below the costal margin. The flat note extends to the pubes. In the erect posture there is tympany in the epigastric and left hypochondriac regions, and on the left side beneath the lower ribs.

The material removed at the last tapping was turbid, somewhat creamy looking. On microscopical examination it contains numerous leucocytes in a condition of disintegration, larger cells in a state of advanced fatty degeneration, and very numerous cholesterin crystals.

Of course the first thought in this case that suggested itself was that the condition was an anomalous form of ascites, due either to chronic peritonitis or to liver disease. The patient's history, the absence of any trace of jaundice, the retention of general nutrition, and the absence of any evidences on examination of enlargement or contraction of the liver seems definitely to rule out hepatic disease—cirrhosis, perihepatitis, or tumor. Nor did it seem likely that any of the known forms of disease of the peritonæum itself could cause recurring ascites without serious deterioration in the health, tuberculosis, cancer, or those remarkable forms of epithelial papillomata involving chiefly the omentum to which Lawson Tait refers.

On the other hand, the existence at first of a bloody

fluid, the filling of the abdomen repeatedly without serious damage to the health, the readiness with which the patient gets up and goes about immediately after the tapping, the physical examination, the presence of cholesterin in the fluid, suggest strongly the existence of cystic disease, either of the omentum or of the pancreas, most probably of the former.

January 26, 1893.—Patient looks in robust health. Abdomen full, but not so much as at former visit. Everywhere soft, but more resistant on the right side; and on deep pressure to the right of and a little above the navel there is an ill-defined mass. No definite fluctuation obtained. Percussion is clear in epigastric and upper umbilical region and to the left side shading off toward the middle of Poupart's ligament. Dull in hypogastric, right iliac, right lumbar, and right half of umbilical, and when he turns on the *left side* the dullness persists and the bowel tympany on the left side is exaggerated.

Tapped again April 8, 1893, and only fifteen pounds of fluid removed. He was tapped June 17th, three fourths of a gallon removed; October 1st, a gallon and a half; and November 26th, one gallon removed.

Additional Note, February, 1894.—He had been doing very well, and had been tapped only once since last date. One Friday, during a business trip, he felt very ill on the train, and had a great deal of abdominal pain and vomiting that night. There was no very special distention of the abdomen, but there was a great deal of sensitiveness on palpation. Attempts were made in various ways to move the bowels without any effect. On Saturday he was very much worse, and seemed collapsed and feeble. There were a few small discharges from the bowels, chiefly of blood. There was not very great abdominal distention. He had vomiting and great depression, which increased, and he died on the Sunday evening.

Through the kindness of Dr. Loeling and Dr. Miller I was notified of the post-mortem and was present. The body was that of a large-framed, well-nourished man. The panniculus over the abdomen was at least two inches and a half in thickness. There was no special distention of the abdomen.

Peritonæum: No exudation, serous or fibrinous. In the right iliac and lumbar regions there was a large cyst, the anterior wall of which was adherent in several places to the abdominal wall, and in addition there were several strong bridles of adhesions, one, the longest, from the left side of the cyst to the peritonæum in the neighborhood of the left crest of the ilium. Two groups of fibrous bands passed from the left cornu of the cyst to the abdominal wall, just to the left of the navel. There were also one or two smaller bands of adhesions, and at one point the upper part of the jejunum was closely adherent to the top of the cyst. After freely exposing the peritonæum by a crucial incision the cyst was seen occupying the position already mentioned. In the lumbar region there were several coils of the small intestine which had passed beneath the bands of adhesion, uniting the left cornu of the cyst with the abdominal wall near the navel. There were two different loops through which the coils of intestine had passed; one anterior, through which about eight inches of the jejunum had passed and the intestine was only slightly reddened, whereas through the posterior loop about eighteen inches of the upper part of the ileum had passed and had become strangulated. The coils were of a deep maroon color, swollen and infiltrated, and the attached portion of the mesentery was enormously thickened and also plum-colored. The peritonæum over these strangulated coils was smooth, there was no fibrinous exudate, and they could be withdrawn without any difficulty through the snare.

The intestines were then removed as far as a foot above the ileocæcal valve. They presented nothing of note except in the strangulated portion just described. After their removal the position of the cyst could be clearly determined. It was of about the size of a man's head, occupied the right iliac and right lumbar regions, and extended to the left beyond the middle line. It lay directly upon the spine and on the lumbar and iliac muscles on the right side. The hand could be placed beneath it, and it could be lifted readily from its bed. The lower foot of the ileum was closely attached to its left and lower margin; below was the cæcum, and the appendix formed a long, flattened, cord-like structure passing up its posterior wall. The ascending colon lay along its right side. The tumor lay in the mesentery of the last foot or eighteen inches of the ileum, and it was removed very readily by stripping the ascending colon and cæcum from its attachment to the peritonæum. The sac had a grayish-white appearance, except at its left side, where it was stained of a greenish color. There were no other adhesions except those mentioned to the parietal peritonæum and to the jejunum. It was a little roughened and puckered in places. When it was laid open, the fluid was a little turbid and slightly blood-stained.

(b) *Multiple Tumor Masses in Abdomen; Phantom Tumor.* CASE LI.—Mrs. E., aged fifty-two years, admitted June 8, 1893, complaining of a swelling below the right jaw, pain in the abdomen, and of having vomited blood. There is nothing of note in her family history.

She has had eight children; four are living. She has never been a very strong woman. Fourteen years ago she had a submucous uterine fibroid, which was removed. Twenty-eight years ago she had a severe attack of typhoid fever. In October of 1891 she had diphtheria, which was followed, she says, by a lump at the angle of the right jaw. Shortly after this, too, she began to have occasional pains in the abdomen. In October, 1892, after a period of a good deal of excitement and worry, she had what was called brain fever, and was unconscious for two weeks. She was in bed at this time for nearly three months, partly in consequence of a carbuncle on the back.

She dates her present illness from about February of this year, when she had a great deal of oppression after eating, sometimes nausea, which were unusual symptoms. She also had at times straining at stool and a desire to go very frequently. The dyspeptic symptoms increased, though she never had severe pain. Four weeks ago, after a day or two of much dyspeptic trouble, she had an attack of vomiting after breakfast, and brought up a large amount of black, clotted blood. Her appetite has been poor, and she has had a great deal of eructation and pain after eating, relieved sometimes by the use of soda. For a week before her admission the right foot began to swell. Within the past six months a tumor mass has developed below the right jaw. Her condition on admission was as follows : Much emaciated; sallow complexion. Attached in front of the angle of the right lower maxilla is a group of enlarged glands, which extend over the jawbone on the cheek. The whole mass can be readily moved; the skin is a little reddened; the individual glands can be felt. The supraclavicular glands are not enlarged. The abdomen looks a little full. The right epigastric vein is much distended with blood; the left vein not quite so large. The current in both is from below upward. In the epigastric region to the left of the navel several nodular masses can be seen beneath the skin. There is no peristalsis visible. On palpation to the left of the navel, there is a solid, somewhat cord-like mass, about six centimetres in length, which extends in an oblique direction toward the axilla. It is very firm and hard, superficial, and feels as if attached to something beneath, as it is only partially movable. No gas is felt bubbling in it. Just above and to the right of the navel is a firm mass which is more difficult to limit and define. Midway between the navel and the ensiform cartilage is a soft, button-like mass which at intervals projects beneath the skin, and then suddenly relaxes with a sizzling sound. In a few moments it appears, hardens into an ovoid, resistant body about three centimetres in lateral extent, and then relaxes again with a sound of gas bubbling in it. The right inguinal region is occupied by a large nodulated mass feeling like a collection of lymph glands. The left inguinal glands are somewhat enlarged. She had no diarrhœa; the stools were liquid, grayish-brown in color. The rectal examination was negative.

The patient remained under observation for two weeks, and no essential change took place in the condition. There was no dilatation of the stomach. The tumor masses above described were very evident.

She left the hospital on June 30th, unimproved.

In many respects no case in the series was more interesting than this one, but a definite conclusion as to the seat of the primary disease did not seem possible. Naturally, with dyspepsia, belching, loss of appetite, and progressive emaciation, one suspected the stomach to be the seat of the malady, the more so with an account of an attack of vomiting in which she brought up a large amount of clotted blood. While under observation in hospital, the condition was not such as to justify putting her to the worry of a test breakfast. Not one of the tumors in the abdomen was apparently connected with the stomach itself, nor were there signs of dilatation. The tumors were rather like masses of enlarged lymph glands. The enlargement of the glands in the right side of the neck dated, she insists, from the diphtheria in October, 1891, but they had enlarged very much since February. The supraclavicular glands were not especially enlarged. Altogether we inclined to the opinion that there was a new growth in the stomach with extensive secondary lymphatic infection. By far the most striking feature in the case was the phantom tumor appearing and disappearing midway between the navel and the ensiform cartilage. Every minute or two it would emerge beneath the skin like a button, get firm and hard, assume an ovoid shape, and then, as one watched it, relax and disappear with a sizzling noise, which could be heard as well as felt. Of course such a tumor is only felt in connection with the tubular muscle of the gastro-intestinal canal, and in this case it was in all probability a limited portion of the coat hypertrophied on the proximal side of constriction, caused either by a new growth at the attachment of the mesentery, or by some narrowing neoplasm in the wall itself. There was admitted yesterday to the ward a case in which you can study another remarkable phantom muscle tumor. The young man has a well-marked history of ulcer, with vomiting of blood and hyperacidity of the gastric juice. The stomach is somewhat dilated, and in the epigastric region there appears at intervals, readily seen beneath the skin, an ovoid tumor, four to five centimetres in length, which lifts the abdominal wall definitely, and then in a few moments relaxes and disappears. When visible it is very firm and hard, and when relaxed it can only just be felt.

(c) *Uterine Fibroids.*—It speaks well for the differentiation of the cases in the hospital that, so far as I know, the following is the only instance of tumor associated with the female pelvic organs which came before me for examination. This, too, was rather by accident. As she had tuberculosis of the lip and tongue, the question was raised whether the abdominal condition was not due also to tuberculosis:

CASE LII. *Tuberculosis of the Lip and Cheek; Multiple Tumors in the Abdomen.*—The patient, aged about forty years, was admitted for tuberculosis of the lip, tongue, and cheek. The abdomen was distended, and I was asked to see her to determine the nature of the masses which could be felt. She stated that they had been present for many years, and had never given her any trouble. The characters were very definite. The lumbar, iliac, the greater part of the umbilical, and the entire hypogastric regions were occupied by solid masses, which in the iliac regions presented several rounded movable prominences. The uterus was firmly fixed, and the whole pelvis appeared blocked with the masses. One point only was of interest in connection with the differential diagnosis of tuberculous peritonitis. In the iliac regions palpation was much softer, and the areas of resistance were separated by distinct intervals, and only on deep pressure could solid, uniform masses be felt. On percussion, there was a flat tympany, such as one finds not infrequently when tuberculous tumors are scattered about among the coils of intestines. Here the history, the character of the masses, and the persistence for more than a dozen years were quite sufficient for the diagnosis.

(*d*) *Sarcoma of the Abdominal Wall.*—The following case is uncommon in my experience and is worth placing on record. Not only was there a large, massive, subcutaneous tumor in the lower umbilical and upper hypogastric regions, but there were secondary nodules beneath the skin of the other parts of the body.

CASE LIII. *Sarcoma of the Abdominal Wall; Numerous Subcutaneous Metastases.*—Mrs. A., aged fifty-two years. The patient had been a healthy woman, had worked hard, and brought up a large family. For two months previous to my consultation with Dr. Atherton, of Toronto, August 24th, she had suffered with diarrhœa and attacks of dyspepsia. The frequent movements have persisted. There was considerable nausea during last month, and vomiting occurred frequently after eating or drinking. She said she felt full, and could not hold much in the stomach. Within the past two months she had lost flesh rapidly and had become very weak. When first seen by Dr. Atherton, August 9th, the tumor about to be described was present. The examination by him of the pelvic viscera was negative. There was no cancer of the breast.

The patient was a medium-sized woman; looked emaciated, scarcely cachectic; was somewhat pale; tongue furred. Nothing of special moment about the circulatory or respiratory systems. The pulse was a little rapid, but of fair volume, somewhat irregular. Slight cough and some bloody expectoration, which continued until her death.

On exposing the abdomen there was seen a remarkable condition of the abdominal walls. While the portion above the navel looked normal, below this point there was a large mass occupying the lower umbilical, the whole of the hypogastric region, and extending into the inguinal regions. The skin was not discolored and showed the lineæ albicantes.

On superficial inspection it might have been taken for a somewhat unusual localized development of the fatty panniculus. On palpation the mass was felt to be firm, hard, involving the skin (which could not be moved separately), and presented a curious nodular feel, suggestive rather of bunches of lymph glands, or of the sensation given by touching the lobulated kidney of a sheep. With this there was also a feeling of massiveness and solidity, as if the tumor extended through the subcutaneous tissues. The mass sloped, as it were, gradually toward the periphery, and here the nodules were more isolated, and in lines running obliquely toward the false ribs. There were chains of these little nodules, like lymph-knots. The whole surface of the body seemed tender and painful on palpation. She complained bitterly of pain after much handling of the mass. There were nodules also beneath the skin of the right breast, one or two in the right thigh, and several in other parts.

Dr. Atherton writes that he saw her for the last time on September 3d, and that she died on September 17th. The daughter states that diarrhœa and vomiting continued, but no blood was passed at any time. For a few days before death "blue lumps," of about the size of a small bean, appeared in various parts of the trunk and extremities, which were tender and painful, like those seen by us.

Whether or not this growth was primary in the abdominal walls is impossible to say in the absence of details which could be furnished by the autopsy alone. She had had gastro-intestinal trouble, but not such as pointed to malignant disease of stomach or bowel.

(e) *Tumors of Doubtful Nature.*—In none of the following cases did it seem possible to arrive at a satisfactory diagnosis. The salient points are as follows:

CASE LIV. *Dysentery; Two Tumor Masses in Abdomen.*— Samuel T., aged about forty years, seen with Dr. F. R. Smith, November 17, 1892, complaining of a lump in the abdomen and diarrhœa. His general health has been excellent, and his family history is good. He had typhoid fever when nine years of age. Ten years ago he was caught between two cars and injured about the hips and legs. He was laid up at this time for four months and had a dull pain in the right side, which has recurred at intervals, but never prevented him from working. Two months ago he appears to have had an attack of acute dysentery, frequent passages, and swelling of the legs. He was in bed for six weeks. As late as October 17th of this year he passed blood in the stools. He now has no swelling in the legs; appetite is good; bowels regular; and he has gained in weight and looks well, perhaps a little sallow. He has had no fever; tongue is clean. No enlargement of the lymph glands. In the right upper quadrant of the umbilical region, extending into the adjacent hypochondriac and lumbar regions, there is a firm, resistant mass, which reaches to about two centimetres below the level of the navel, and to the right can be felt as far out as the tip of the tenth rib. To the left it does not quite reach to the middle line. On bimanual palpation it is freely movable. It is not well felt below the ribs behind, but on deep pressure it can be pushed forward, and then is distinctly movable. It can be readily separated above from the liver, the edge of which is easily to be felt. In the hypogastric region, just above the pubes, and to the right of the middle line, there is a second tumor, feeling about the size of an orange, somewhat elongated. To the right it appears to have a definite ridge-like edge. The percussion is everywhere resonant, and a clear note is elicited over both tumors, except for a short space to the right of the middle line over the one in the hypogastric region. Neither spleen nor liver dullness is increased. There is no dilatation of the stomach. The urine is clear and he has passed no blood, and there is no albumin; no tube casts.

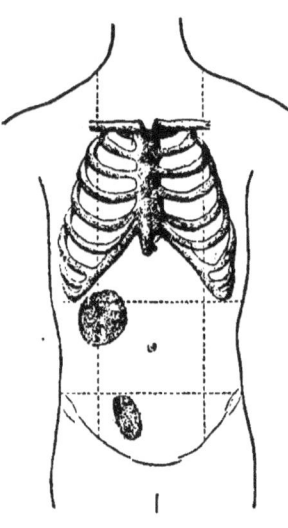

Fig. 37.—Situation of the tumors in Case LIV.

The patient was seen again January 10th. General condition has been excellent, and he has been at work. There is no cachexia. The tumor masses previously noted were present with great distinctness. The lower one appeared to be scarcely so large as on previous examination. In the left inguinal region, about three centimetres from Poupart's ligament, there is a freely movable, subcutaneous nodular body about the size of a bean. The inguinal glands are not enlarged.

It did not seem possible to arrive at any conclusion as to the nature of these tumors. The strongest possibility seemed to be of their connection with the intestine, as he had tenesmus, diarrhœa, and melæna, but the patient's excellent general condition and rapid improvement would seem to contraindicate new growth.

While preparing this lecture for the press this patient was seen (December 20th). His condition remains excellent and he has been steadily at work. The lower tumor is no longer palpable; the upper is still quite distinct, though smaller than at date of former note. Its position is unchanged.

The following case I regarded at first as tumor of the stomach, but on subsequent examination it seems to be extremely doubtful whether it is really in this organ:

CASE LV. *Tumor Mass in the Epigastrium of Doubtful Nature.*—Kate H., aged forty-one years, admitted August 29, 1893, complaining of pain in the left side and swelling in the epigastrium. There is nothing of moment in the family history. She has been healthy with the exception of typhoid fever at twenty-one and pleurisy last winter. She has had at times irregular cramps in the abdomen. For several years she has had dyspepsia, suffering with belching after eating and slight discomfort. Ten months ago she had an attack of severe pain in the epigastrium, with nausea and vomiting, and was in bed for a week. In the next two months she had four similar attacks, which lasted about a week. Some of these attacks, the doctor said, were due to gallstones. About two months ago she first noticed the swelling in the abdomen which is now present. She does not think it has increased in size. She never has been jaundiced; bowels are regular; passes a normal quantity of urine. Her chief complaint is of a dull pain below the left costal margin. She is a fairly well-nourished woman; lips and mucous membranes are pale; tongue is coated. The abdomen is distended, chiefly in the epigastrium and in the right hypochondriac region, in which there is a smooth prominence. On palpation this corresponds to a rounded, hardened, somewhat nodular tumor, which in the middle line feels quite smooth and on the right side is more irregular. In the nipple line there is a marked ridge to be felt midway between the costal margin and the transverse navel line. The liver dullness begins at the sixth rib, and continues directly into this prominent tumor mass. At first it was thought that this represented an enlarged liver, but on September 2d, on palpation, gurgling sounds were noted " in the prominent tumor in the epigastrium, and to-day it is everywhere resonant. The gurgling can be felt in this solid mass with the greatest distinctness. The tympany reaches as high as the base of the ensiform cartilage." Ewald's test breakfast showed the presence of free hydrochloric acid. This case interested us extremely from the remarkable simulation of the outline of the tumor mass to that of an enlarged liver.

On September 5th the following note was made: "There is a very distinct prominence below the costal margin in right epigastric and right hypochondriac regions. On palpation, a firm, resistant mass fills the upper zone of the abdomen, the outline of which resembles quite accurately that of an enlarged liver. Beyond the left parasternal line no very distinct edge is to be felt, but toward the right, as she draws a deep breath, there is a distinct nodular edge. The entire mass descends with inspiration. The resonance is not so extensive to-day; it does not reach beyond the right parasternal line. After inflation the stomach tympany extends two fingers' breadth below the navel, and there is in the left epigastric and the left upper quadrant of the umbilical region the outline of a dilated stomach.

When resonance was first noticed in the mass below the ensiform cartilage we thought that possibly extensive infiltration of the stomach wall existed; but the patient has been under observation on several occasions since, and she has gained in weight, looks well, free hydrochloric acid is present in the gastric juice, and after inflation the resonance is not more tympanitic, and the stomach outlines seem somewhat below the mass. Altogether there was doubt enough to exclude the case from the stomach list, and I place it here among the miscellaneous tumors of doubtful nature.

(*f*) *Aneurysm of the Aorta.*—And lastly, not the least interesting of the miscellaneous tumors was a large sacculated aneurysm of the abdominal aorta.

CASE LVI.—Lee K., aged sixty-seven years, admitted July 5th, complaining of a " fluttering lump" in the abdomen. With the exception of scurvy and rheumatism during the civil war he has been a very healthy man. He is temperate and denies venereal disease, but there is a distinct cicatrix just beyond the glans penis.

Present Illness.—For three years he has noticed a lump in the abdomen, which for the past two years has been painful, and which he says has lately increased in size. He has a dull, steady, gnawing pain in the tumor itself, and more or less pain in the back. The pain and throbbing sometimes nauseate him, particularly after eating, and he has vomited twice in the past two weeks. Bowels are constipated; he has severe headaches, and has had occasional bleeding from the nose. He is short of breath on exertion. Patient is a very vigorous, healthy-looking man; well built; well nourished; musculature above the average. The conjunctivæ are a little watery and yellow; pupils are equal. Pulse regular, equal in both radials; the arteries thickened; can be rolled under the finger; pulse wave can not be obliterated. The brachial arteries are tortuous. The examination of the lungs is negative. The apex beat of the heart is seen in the fifth interspace, just outside the nipple line; it is forcible and well defined. The sounds are clear; the first a little thudding, and both somewhat accentuated at the base.

Abdomen.—In the epigastric and upper part of the umbilical regions there is an irregularly rounded prominence, which pulsates forcibly and almost synchronously with the heart impulse. It has a transverse diameter of 8·5 centimetres; vertical nearly eight centimetres. It is perhaps a little more prominent to the right than to the left of the middle line, and on the right side almost obliterates the groove below the right costal margin. On palpation it feels smooth, yields to firm pressure, expands forcibly and in all directions. There is at times a distinct systolic thrill. The borders are everywhere rounded and it seems to dip down rather sharply just above the umbilicus. The tumor is not influenced by the knee-elbow position. The whole mass can be grasped in the hand and the expansion in that way very readily felt. It is unusually mobile laterally. It can be moved to the right, so that its left border is at the middle line. It can not be moved to the left quite so far, but far enough so that it pulsates under the left costal margin. The up and down movements are very slight; it is not influenced by respiration. There is dullness on light percussion over the top of the mass, and in a circle the diameter of which would be five centimetres. Beyond this there is tympany on all sides. On auscultation there is a loud systolic bruit heard everywhere over the tumor. There is no definite diastolic shock. The condition of the liver, spleen, and other parts of the abdomen is negative. The diagnosis of aneurysm of the abdominal aorta was made by Dr. Hewetson, under whose care the patient first came, and subsequently when I saw him the doubt arose in my mind, owing to the extreme mobility, whether it was really in the aorta, or whether it might not be connected with one of the branches—the tumor seemed remarkably mobile, and could be pushed so far from left to right. Dr. Halsted, too, thought that the tumor might possibly be in one of the branches; and as the patient consented he did an exploratory operation. The tumor was found to spring directly from the aorta just above the renal arteries. The pedicle of the sac was short and almost as wide as the aneurysm itself. It was thought better to leave the case to Nature than to attempt any measures to promote consolidation in the sac. The patient recovered rapidly from the operation and left the hospital in about ten days.

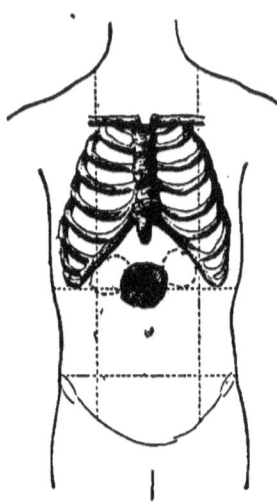

FIG. 38.—Position of the aneurysm. The dotted outlines illustrate the extreme mobility. Case LVI.

Aneurysm of the abdominal aorta is rare. This is the first one which has been under our observation since the hospital was opened, during which time there have been between forty and fifty aneurysms of the thoracic aorta in the wards. The diagnosis here was readily made; the tumor was so pronounced, so rotund, so expansile in all directions, and with a well-marked thrill and systolic bruit—no single feature of aneurysm was absent. The mobility alone was unusual; not one of the few aneurysms in this situation which I have seen presented such remarkable mobility.

A few weeks subsequently I saw in Montreal with Dr. Shepherd a patient who had progressive anæmia and debility with great abdominal distention and pain. An abdominal tumor had been suspected, but the tympany and distention of the stomach and bowels prevented any satisfactory examination. She became more anæmic and died the day after I saw her. Through the kindness of Dr. Wyatt Johnston, I saw the specimen, which proved to be a large aneurysm of the abdominal aorta which had compressed the duodenum, causing great dilatation of the stomach. It had ruptured at one edge and hæmorrhage had taken place into the retroperitoneal tissues.

LECTURE VI.

TUMORS OF THE KIDNEY.*

NOWHERE is the close interdependence of medicine and surgery better illustrated than in the diagnosis and treatment of tumors of the kidney. A very large proportion of the cases come first under the care of the physician, whose province it is to recognize the condition; but to do justice to his patient he should be thoroughly familiar with the advances which have been made in the department of renal surgery. Let me first call your attention to the diagnosis of certain conditions associated with—

1. MOVABLE KIDNEY. (a) *Errors in the Diagnosis of Movable Kidney.*—I have no statistics to offer with reference to the frequency of movable kidney, but throughout the session you have had many opportunities of noting its commonness—so common, indeed, that we are never without examples in the wards. A majority of the cases present no symptoms whatever. Others complain much of dragging pains in the back, with neuralgia, epigastric distress, and general nervousness; many of neurasthenia with dyspepsia; and one often finds, particularly in women who have borne children, the condition to which Glénard has given the name enteroptosis. In a thin person, male or female, who presents the general features of neurasthenia, you will be almost certain to find, on examination, mobility of one or other, or of both the kidneys. Inability to

---

* Concluding lecture of the course. Delivered Dec. 26, 1893.

lie comfortably on the left side, and paroxysmal attacks of pain such as I shall describe in a few moments, are less frequent symptoms. It is difficult really to determine how far all these features are dependent on the renal condition. It is quite possible that the pains and uneasy feelings may be due to stretching and tension of the tissues in the neighborhood of the great abdominal nerve plexuses; but one not infrequently meets with cases of the most extreme mobility without any symptoms whatever. It may be that, in persons with a debilitated and bankrupt nervous system, the tension caused by the dragging of a movable kidney may be at once felt, just as many persons find the first indication of physical fagging in subjective sensations of the movement of the heart, of which in health we are not cognizant. The text-books and monographs now contain a full and satisfactory account of the condition, but I wish to call your attention to some less widely recognized features in connection with it.

While in the great majority of all cases movable kidney is quite unmistakable, there are cases in which its recognition is by no means easy. You will remember that in Case V we made a somewhat serious *faux pas*, and thought that an unusually mobile pyloric tumor was a movable kidney. A more frequent error is the mistaking of it for a dilated gall bladder. I have already alluded to this, and in *Lecture IV* have spoken of the points to be attended to in the diagnosis. Here I may mention a case of a good deal of interest in which this error was made, and the operation for dilated gall bladder performed by Mr. Tait. The patient, a doctor from California, consulted me in 1888 about a lump in the abdomen, the nature of which had puzzled a large number of physicians. It had been present for ten or eleven years, and had appeared first after a somewhat severe attack of sea-sickness. He had suffered a great deal with nervous troubles and dyspepsia, and when I saw him there were signs of dilatation of the stomach. The position, mobility, and general characters made me feel tolerably certain that the tumor was a movable right kidney, and with this opinion the late Dr. Agnew coincided. The diagnosis relieved his mind very much, and he improved, so far as nervous symptoms were concerned. Subsequently he grew worse, and in the spring of 1890 he consulted Mr. Tait, who diagnosticated dilated gall bladder, and made an exploratory incision. The gall bladder was normal, and the kidney, so he stated, was *in situ*. Subsequently the patient came under the care of Dr. C. O. Baker, of Auburn, N. Y., who found the tumor very evident, made a diagnosis of floating kidney, and performed nephrorrhaphy, with great relief to the patient. The case is reported in the *Medical Record* for May 14, 1892. The patient died in September of this year (1893) of empyema, and in the post-mortem notes, which were very kindly sent to me, the statement about the kidney is: "Right organ in normal position, held by firm union; nephrorrhaphy had been performed for floating kidney." There is no mention made of enlargement of the gall bladder.

With very pendulous and lax abdominal walls and an unusually mobile right kidney there may be at first difficulty in separating clearly the right lobe of the liver and the kidney. In the following case I was at first in doubt:

CASE LVII. *Movable Kidney, simulating a Local Growth in the Right Flank ; Right Lobe of Liver mistaken for Right Kidney.*—Jane E. G., aged fifty-two years, seen with Dr. Hewetson, in the Medical Dispensary, September 25, 1892, complaining of cough and pains in the back and side and headache. She is pale and somewhat emaciated, and looks ill. Lips and mucous membranes are, however, of a good red color. She has borne nine children. She has lost in weight during the past six months. There are no abnormal physical signs in the thorax. In the examination of the abdomen there was detected a solid, mobile tumor in the right side, which was not thought to be the kidney, as it was believed that this organ could be also felt in the flank. On the examination I found a well-marked, readily movable tumor, just at the right of the navel, and, on bimanual palpation, the right kidney could, I thought, also be felt. The left kidney was readily palpable. As I was unable to satisfy myself as to the nature of the mass, I asked to see her again on the 28th.

Examination.—Patient is thin; not cachectic. The abdomen is pendulous, and the walls are very lax. On the right side, a little above the line drawn from the anterior superior spine to the navel, a movable tumor can be felt, somewhat rounded in shape, about the size of an orange, not distinctly reniform, but with a slight depression on the right side. It can be pushed up, but not entirely, beneath the ribs, and it does not slip into position like a floating kidney. Below it can be pushed down so as partially to pass the line joining the anterior superior spines. To the left it can be pushed over to the middle line. It is not painful. Examination of the right renal region showed a depression below the ribs behind, but, on bimanual palpation, the flank appeared to be filled with a solid mass. Careful examination, however, determined that this was not, as suspected on the first examination, the kidney, but in reality part of the right lobe of the liver, the ligaments of which were much relaxed. An edge could be distinctly felt; to the right the lobe was ill-defined, and it could not be made to slip up in the way so characteristic of movable kidney. On the left side the kidney was readily palpable, and, on deep inspiration, depressed so much that the fingers could almost be inserted above the upper border.

I have no doubt that in this case the tumor in the right side of the abdomen was the kidney, and that the mass felt in the right flank represented in reality the right lobe of the liver.

Among the scores of cases of movable kidney which have come under my observation I do not remember one in which the condition was exactly as presented in this patient. At the first examination I felt sure that a kidney was palpable in the renal region, but the more careful subsequent examination convinced me that I had been in error.

I have already called your attention to the elongation of the edge of the right lobe of the liver as a cause of an anomalous tumor mass in the right flank, and mentioned the case in which laparotomy was performed by Dr. Kelly for an obscure tumor, which proved to be the thinned-out edge of the right lobe.

In the following case I erred in thinking that a tumor in the right side of the abdomen was a dislocated and fixed kidney:

CASE LVIII. *Tumor in Right Side supposed to be a Dislocated and Fixed Kidney; Gradual Disappearance.*—Mrs. H. O., seen December 9, 1892, with Dr. Arthur Williams, of Elkridge, Md.

I saw this patient first in October, 1891, when she consulted me for pains in the abdomen. She had always been in good health as a girl; had been married twelve years; had four children, the youngest three years of age. Six months ago she began to have pain in the abdomen, chiefly in the epigastric region, and radiating to the back and to the chest. She has never had any vomiting, nor is the pain connected with the taking of food. Within this period of time she has lost about sixteen pounds in weight. A few months ago she noticed a lump in the abdomen, which has caused her great uneasiness, and it was for the purpose of determining the nature of this that Dr. Williams advised a consultation.

Examination.—She was a thin, dark-complexioned woman, neither anæmic nor cachectic. The abdomen was distended, uniform, with normal respiratory movements. On palpation not tender, not sensitive, and nothing was felt except at the boundary of the epigastric and umbilical regions on the right side, where an elongated tumor occupied exactly the extension of the parasternal line. The lower end was rounded and smooth, and reached a little below the level of the navel. The upper end was not palpable. To the left the mass did not extend beyond the middle line. The right margin was rounded and well-defined; the left a little depressed and irregular. It was not very movable, but by using both hands it could be shifted slightly from side to side. The surface was smooth ; it was a little sensitive to pressure, and felt very resistant and solid. The fingers could be placed directly beneath the lower end, but it did not appear to have an ovoid or globular outline. No gas was felt to pass through it after repeated examinations. The stomach was not dilated. The kidney could not be felt on either side. The urine was normal and the bowels were regular. She had no attacks of colic, and the pains which she described were not specially suggestive of biliary colic. I noted at the time that the case was somewhat puzzling. The tumor had the situation rather of a pyloric growth, though its long axis was vertical, but there had been no dyspepsia, and there was no dilatation of the stomach. The situation was somewhat suggestive of the gall bladder, though it seemed to have more resistance than a tumor caused by dilatation of this organ. It had an outline very suggestive of renal tumor, and I was rather inclined to regard it as a dislocated and fixed kidney. I gave the patient every encouragement and assured her that it was not a malignant growth. At my request the patient returned on December 9, 1892. To my astonishment the tumor had disappeared entirely ; not a trace of it could be felt. Examination of the abdomen was absolutely negative. The right kidney was not palpable ; the gall bladder could not be felt ; the edge of the liver could be just touched during deep inspiration. The patient stated that she had improved very much. The pains had diminished, and she now had very little distress. She had gained in weight, and was on the whole very much better. Here, in all probability, I had mistaken a dilated gall bladder for a movable kidney. The tumor was scarcely large enough, and had not the situation of intermittent hydronephrosis, the only other one which could disappear in this way.

(b) *Dietl's Crises in Movable Kidney.*—Remarkable attacks of pain occur in connection with movable kidney, to which attention was first called by Dietl. A knowledge of the existence of these renal crises, as they have been termed, is very important, and as they form a very striking feature in certain cases of movable kidney, I propose to call your attention to them at some length. The textbooks, with the exception of the last edition of Flint's, have been curiously silent regarding this symptom group. In Dietl's paper, which appeared in the *Wiener medicinische Wochenschrift*, 1864, nine cases of movable kidney are reported, all of which had pains in the side and back. In four there were also attacks of nausea and vomiting, with great pain, swelling, and tenderness of the affected kidney. These were liable to recur, particularly on exertion. Dietl was doubtful about the pathology of the condition, but from the title of the paper, Wandernde Nieren und deren Einklemmung, it is evident that he regarded it as a strangulation caused probably by a twist in the vessels.

The case which first called my attention to the condition was a patient of the late Dr. Palmer Howard's, of Montreal.

A lady, aged about forty years, stout and well nourished, began some months after her third pregnancy to have violent attacks of pain in the abdomen, in which she became nauseated, often vomited, and suffered so intensely that hypodermics of morphine alone gave relief. She was seen by the late Dr. George W. Campbell, who discovered a lump in the right side of the abdomen. The attacks recurred with great severity throughout the winter of 1879-'80. The patient lost in weight and the diagnosis of a new growth was made. In the spring of 1880 she consulted in New York the late Dr. Austin Flint, who agreed with Dr. Howard and Dr. Campbell as to the very serious nature of the case. Throughout the year the attacks recurred, and she lost in weight from one hundred and seventy to one hundred and twenty pounds. In the spring of 1881 she again went to New York and consulted Flint and Van Buren. As she was at this time very much thinner, a more satisfactory examination of the abdomen could be made. Van Buren at once suggested that the tumor was a movable kidney, with which he stated that he had frequently met with paroxysmal attacks of severe pain, particularly in gouty persons. He advised a very strict diet. The relief of mind was naturally very great, and the patient began at once to improve, gaining rapidly in weight. The paroxysms reduced in frequency, and for years she remained well, having at intervals, particularly if she committed any indiscretion in diet, recurrences of the severe pain.

The following cases have been recently under observation:

CASE LIX. *Enteroptosis; Movable Right Kidney; Severe Renal Crises.*—Susan S., aged forty-six years, admitted January 13, 1893, complaining of agonizing pain in the abdomen and back, and a lump in the right side. She was married at twenty-three ; has had nine children, no miscarriages ; menopause two years ago. Of late years she has been very nervous and is often irritable and depressed. At the time of the menopause she had pains in the back, and once the head was drawn to one side for a few days. Until two years ago the pains were of a dull aching character, but at this time she noticed a lump in the right side, and the pains became much more intense. The attacks now come on without warning and are so severe that she becomes helpless. They last for several hours, and though she never loses consciousness, they are so agonizing that for a time she can not speak. On two or three occasions she has fallen down. She gets cold, sweats a good deal, feels nauseated, but has never vomited. The pain is chiefly in the right side, and the lump becomes, she says, sensitive and larger. The attacks have recurred every two or three months. The last one was four weeks ago. On several occasions after very severe attacks the urine has been dark-colored. She has never had jaundice. The patient is a well-nourished woman ; lips and mucous membranes of good color ; temperature normal ; examination of heart and lungs negative.

The abdominal walls are greatly relaxed, and the much-scarred skin can be grasped in large folds. On the left side the kidney can be felt readily on deep inspiration. On the right side, extending outward to within 3·5 centimetres of the middle line, and downward at least 8·5 centimetres from the costal margin, is a smooth, rounded mass, very freely movable to the right. It is superficial and seems to emerge directly beneath the ribs. It descends with inspiration, and when the patient turns on the left side, falls far over beyond the middle line, and can be lifted with the fingers beneath it. It is smooth on the surface, and as stated seems to emerge directly from beneath the costal margin. To the left it can be felt beyond the middle line. The lower edge is rounded, but the fingers can not be placed beneath it. It is evident that this mass is a depressed and somewhat freely movable liver. On bimanual palpation, deep pressure opposite the point of the tenth rib, the right kidney can be readily felt behind and separate from the liver, and on deep inspiration it moves down and can be readly grasped.

We had a good deal of discussion about the nature of this large flat mass in the right flank. It felt very superficial, smooth, and we thought at first it might be an enlarged and movable kidney, but repeated examination seemed to indicate that it was the liver, somewhat movable and tilted forward, owing to the relaxed condition of the abdominal wall. The following note was made on the 19th : When the patient lies on the left side the hepatic flatness does not begin in the anterior axillary line until the ninth rib ; when on her back it begins at the eighth. In the nipple line, when she is on her back, the flatness begins at the seventh rib, and apparently falls an inch lower when she is on her left side. The border of the mass can be felt more distinctly in the nipple line, and suggests, taking into consideration the fact that one feels below it the kidney sliding backward, while near the umbilicus there is a sharp border, the existence both of a movable kidney and a movable liver.

The patient remained in hospital for nearly four weeks and gained in weight. She was greatly relieved by our statement as to the nature of her case. She had no attack of severe pain while under observation. The character of the attacks suggests the renal crises common in floating kidney.

The following case is, I believe, a very typical instance of Dietl's crises, and is interesting also from the protracted course and the intensity of the recurrences :

CASE LX. *Movable Kidney ; Renal Crises at Intervals for Seven Years.*—Dr. X., aged forty-three years, seen October 18,.

1893, complaining of attacks of agonizing pains in the abdomen, which have recurred on and off for seven years. The patient has been a very healthy man, of good habits, and has for twenty years been engaged in a very laborious country practice. At the time of the onset there was a great deal of typhoid fever in his district, the roads were very bad, and for seven weeks he was in the saddle constantly. The first attack, which was of a very agonizing character, came on when he was very much fatigued, and was so severe that he nearly fainted and required morphine. He had no vomiting and did not pass any blood in the urine. Since that time the attacks have recurred, sometimes two or three in a week, sometimes only one in six or eight weeks. He has never vomited in them, though sometimes the intensity of the pain makes him nauseated. The bowels are regular and he has never had jaundice. He never can tell exactly when the attack will come on. It usually begins abruptly and the intensity of the pain is such that he often has to take chloroform. The attacks last from a few hours to the greater part of a day, and, in passing away, leave him a little exhausted and with a feeling of soreness and aching. The pain is most intense in the right flank and extends toward the navel and to the spine. He does not think that any tumor develops at the time, but the muscles of the abdomen are tightly contracted and the right flank is sensitive. He has noticed in very many of the attacks that he micturates freely, and the amount of urine is increased as the attacks pass off. There never has been any change in the character of the urine.

The patient is a moderately well-built man; looks healthy and strong; tongue is clean. The abdomen is soft, flat; no sensitiveness over the stomach; the pylorus is not palpable; the edge of the liver can be felt just below the costal margin; it is not sensitive. The spleen can not be felt.

The right kidney is readily palpable, and, when he draws a deep breath, comes down so low that the fingers can easily be slipped above it and fix it below the level of the tenth rib. It is a little sensitive on pressure, but is not apparently enlarged. When he turns on the left side the kidney falls forward and can be also readily felt just below the margin of the liver. The left kidney is not palpable.

These renal crises constitute perhaps the most distressing symptoms of movable kidney, and they are, I think, very much more common than we are led to suppose. The knowledge of their existence is important, as the attack may be so severe as to simulate peritonitis. The cause of the symptoms is not at all clear. The terms which have been used, *Einklemmung* by the Germans, and *étranglement* by the French, are based upon the view originally expressed by Dietl, that it was a condition of strangulation or extreme engorgement caused by a twist in the vessels of the floating kidney. Dietl thought that about the moving organ there was a local peritonitis. The explanation which passes current at present is more reasonable, namely, that the condition is due to a kink or twist in the upper part of the ureter, with retention of the urine in the pelvis and calices, and a production of a transient hydronephrosis, the severe, agonizing pain being caused by the distention of the tissues.

II. INTERMITTENT HYDRONEPHROSIS.—With the exception of a remarkable case of the rare congenital form, upon which my colleague Halsted operated in this hospital three years ago, I had never seen—to recognize—a case of intermittent hydronephrosis. During the present session three examples have come under my notice. Let me first read to you the notes of the cases:

CASE LXI. *Intermittent Development of Large Tumor on the Left Side.*—Mrs. F., aged forty-three years, seen with Dr. Finney, September 9, 1893, complaining of trouble in the left side. She has been a healthy woman; has had four children; never has had any trouble after her confinements, and she does not think that she was unusually large during her pregnancies. She has, on the whole, enjoyed very good health. In April last she stumbled over a slop jar and wrenched her back, but she did not feel it much at the time. Early in May she had the first severe attack of pain in the left side, which Dr. Archer, who attended her, thought was renal colic.

TUMORS OF THE KIDNEY.

There were three paroxysms—at 3, 6, and 9 P. M. They were evidently very severe, as she was bent over with the pain and had severe vomiting. The urine was not bloody, and in a few days she was herself again, but one evening she was surprised to feel a "lump" in the left side, which has been present at intervals ever since. It has not been especially painful, but is a little uncomfortable, and associated with a feeling of distention and uneasiness, particularly when she is lying down. It is not more painful after eating, nor has diet any special influence. She has not lost in weight. She is quite positive that the lump in the side appears and disappears; thus, she says, she could not feel it on the 5th and 6th of this month, and she thinks that throughout the greater part of July it was not present. Its onset is always ushered in with pain in the left side, but the attacks have never been so severe as those which she had in May. She has noticed on several occasions that she has voided large quantities of urine, as much as five pints between 8 P. M. and six o'clock the next morning, usually of a very pale color. She has not had her attention drawn to any coincidence between the disappearance of the tumor and the large amount of urine.

Her bowels are regular, appetite good, but she has been sleeping badly of late, owing in part to the worry about the tumor.

Present Condition.—Well-nourished, healthy-looking woman of medium height. The abdomen looks natural; no special prominence. When she turns a little on the right side there can then be seen a projection in the left flank just above the ilium, and between the tenth rib and the anterior spine there is felt a prominent solid mass, which above lies close beneath the ribs, while anteriorly it feels superficial. It can be readily grasped between the hands and moved to and fro. When she draws a deep breath

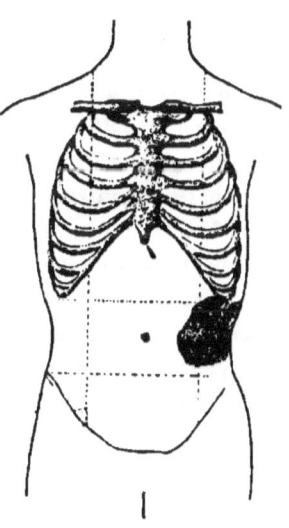

Fig. 39.—Illustrating the position of the tumor in Case LXI.

it does not give one the impression of coming out from beneath the ribs and is not much depressed. No sharp edge can be felt, but it is everywhere rounded in outline.

Percussion in the splenic region is clear, and beneath the level of the eighth rib there is a flat tympany in midaxillary line. As she turns on the right side the mass comes forward and produces a bulging beneath the skin. It is tolerably firm and elastic, but fluctuation can not be obtained.

The edge of the spleen is not palpable; the liver dullness is not increased; the edge can not be felt. The right kidney is just palpable on deep inspiration. Examination of the thoracic viscera is negative.

The patient was requested to make a careful estimation of the urine each day, and note with reference to the presence or absence of the tumor.

September 11th.—Dr. Finney reports that last night on examining the abdomen no trace of the tumor could be felt.

She was ordered a bandage with a carefully adapted pad, and asked to estimate the amount of urine, which she only did, however, for about a week. On the 11th the amount of urine was five pints and a fifth; on the 12th, three pints and a half; on the 13th, four pints and a half; on the 14th, four pints and a half; on the 15th, three pints and a half; on the 16th, two pints and a half. From 6 A. M. to 6 P. M. on the 16th she felt tired and weak, and had uncomfortable sensations, and she passed at this time not quite a pint. At 11 P. M. on the 16th the tumor mass was quite evident, projecting prominently between the ribs and the hip. It was evident throughout the 17th, but she felt very much better toward the afternoon, but was inclined to cry and fret, and was a good deal distressed at the recurrence of the mass. On the 18th it had disappeared entirely. The sample of urine examined was clear, specific gravity 1·015, and contained neither albumin nor tube casts.

Additional Note.—I saw this patient last on January 8th. She had been very nervous and uneasy about herself. The tumor was present, though not so large as when first seen. Its appearance and disappearance have been repeatedly verified by Dr. Finney.

CASE LXII. *Attacks of Colic; Tumor in the Left Renal Re-*

gion which Appears and Disappears.—Mrs. A., aged twenty-seven years, bipara, had consulted me on two or three occasions for dyspepsia. On October 2d she came complaining of a lump in the left side, which had been present on and off all the summer, and which sometimes gave her a great deal of pain. She first noticed it in May, following an attack of colic of great intensity. A few days subsequently she noticed that there was a lump in the left side, which, however, gradually went away. Since then it has appeared and disappeared five or six times, at intervals of a week or two, usually developing with an attack of pain, which gradually subsides as the tumor becomes apparent. No special uneasiness attends its disappearance, and she has not noticed any special increase in the amount of urine.

Her general condition has kept very good; she has gained in weight, and the old dyspeptic symptoms for which I saw her last year have almost entirely disappeared.

She is a medium-sized, fairly well nourished woman; color of lips good. Pulse 80; no fever.

The abdomen looks natural; skin not very much scarred; only a moderate amount of panniculus. A prominence can be seen in the left flank. On quiet breathing, below the left costal border, in the position of the edge of an enlarged spleen, there is a rounded, superficial mass, which descends with inspiration, reaching fully three fingers' breadth from the margin. It passes deeply in the region of the kidney, and the fingers can be inserted between it and the costal margin. Pressure from behind in the left flank pushes the mass forward so as to elevate the skin to the left of the navel. When very large she says it reaches quite as far as the middle line. The lower end is rounded, but without any marked prominence; it is movable when grasped between the hands. The splenic dullness can not be obtained, nor is an edge to be felt. The percussion over the tumor mass is flat. When lying on the left side it falls down somewhat, and the fingers can be passed freely beneath it. There is no uterine or ovarian trouble, and the tumor mass does not appear to pass toward the pelvis. The liver is not enlarged; there is no dilatation of the stomach.

October 7th.—The mass is stated to have been absent for three days.

Examination.—Fingers can be passed deeply at the left costal margin without meeting anything. The kidney is readily palpable and descends with inspiration. She passed three pints and three quarters of urine during the three days after the last examination; then on Thursday and Friday the lump had disappeared, and there were three pints and a quarter, so that the difference is not very great.

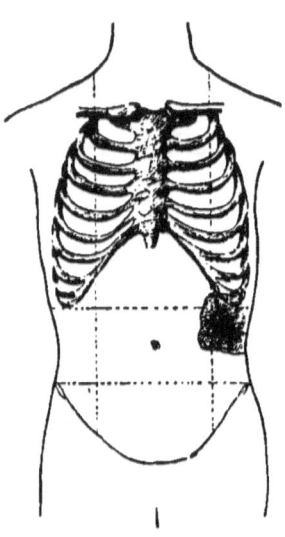

Fig. 40.—Position of the tumor in Case LXII.

The attack in which I first saw the patient came on Sunday, October 1st, with moderately severe pain, and during the night the "lump" was felt. She states that it was not so prominent as it has frequently been. Twice the mass has extended as far as the middle line.

The urine was examined repeatedly. It was light in color; specific gravity never above 1·020, acid in reaction, and contained no albumin or abnormal ingredients. There were no differences between the urine when the tumor was present and that passed during the time of disappearance of the mass.

Throughout November the patient was very well. A carefully adapted pad and bandage have given her great relief, and she has noticed the tumor only once. She has been in Philadelphia staying with friends. Throughout the week ending December 16th she had had a great deal of worry and trouble with illness in the family, and had been on her feet a great deal. On the 15th she was tired out and went to bed. Stayed in bed all day and had some pain and distress on the left side and noticed the reappearance of the tumor.

December 16th.—To-day, at 4.30 P. M., the tumor mass is present, though not so prominent as on the former examination. It does not reach beyond the parasternal line; is not specially sensitive; easily moved on bimanual palpation. On deep inspiration the fingers can be placed well above it.

On questioning the patient with reference to the onset of the attacks, she states that even when a young girl she remembers to have had pain in the left side after running or after dancing for a long time, but she never noticed the presence of the lump until May of this year. She has given up measuring the daily amount of urine, but she is certain that there is no striking and sudden increase in the amount as the tumor disappears.

CASE LXIII. *Pains in Left Side, with Development of Tumor, which gradually Disappears.*—Mrs. X., aged forty-six years, admitted to Ward C, October 23d, complaining of intermittent attacks of pain in the left side, and a swelling or lump which occurs at the same time.

The family history is good. She was always very strong as a girl; married at twenty-two; has had three children, the youngest now ten years old. Was never very large during her pregnancies. Has been always regular; has no uterine disease; still menstruates. Of late years she has had a good deal of mental worry and trouble, and has had a very busy life, actively engaged in housework.

The attacks of which she complains date as far back as eight or nine years ago, and consisted then of pain in the left side occurring once in one or two months, which was, however, quite bearable. It sometimes followed imprudence in diet; sometimes after a jolting ride. The worst attack, shortly after the trouble began, followed a day's journey on the railroad. The pains were never so severe as to require morphine, but there was a sensation of uneasiness and of discomfort and aching in the left side. Nearly four years ago she first noticed a swelling beneath the ribs on the left side. It was not large and usually only lasted a day or two. She can always tell for twenty-four hours before an attack comes on from curious dull, heavy feelings all over her, and then the backache and dragging sensation in the left flank begin. Within the past year or so the attacks have been more frequent, and not a month has passed without them. The lump, too, has become more prominent during the attacks. Lately they have recurred as often as every week, and for the past month they have begun regularly on Sunday. She does not think that any special diet brings them on, nor has she noticed lately that exercise or jolting has any influence.

The urine has been clear; she has not noticed any special difficulty, nor has she had any trouble in micturition. The bowels are sometimes constipated, and more particularly at the time of the attacks.

Patient is thin, weighs only one hundred and five pounds, and is pale. The following note was made at noon of October 24th: The abdomen is flat—not specially scarred. On palpation it is everywhere soft until toward the left costal margin, where a large mass can be felt occupying the left side of the abdomen, and projecting apparently from beneath the ribs. Anteriorly it extends into the umbilical and epigastric regions as far as the middle line. The lower margin, rounded and smooth, is nearly at the level of the anterior superior spine. At first it was thought possible from its situation to be an enlarged and somewhat irregular spleen. It descends with inspiration, and during the deepest breath the hand can be passed over it, and the mass in this way held down. It can readily be felt from behind, and on bimanual palpation can be grasped between the hands, and on firmest pressure below the ribs behind the mass can be pushed forward so as to lift distinctly the abdominal wall. The lower and posterior surfaces appear to be irregular. The sensation given on deep pressure is of an elastic resistance. On percussion there is tympany over the mass in front, a flat tympany in the midaxillary line, and dullness behind. The right kidney is distinctly palpable and descends far enough on inspiration to be held down. For the first twenty hours in hospital patient passed only 380 c. c. of urine, clear, straw-colored; specific gravity, 1·006; slight trace of albumin; no sugar; a few leucocytes, and flakes of epithelium.

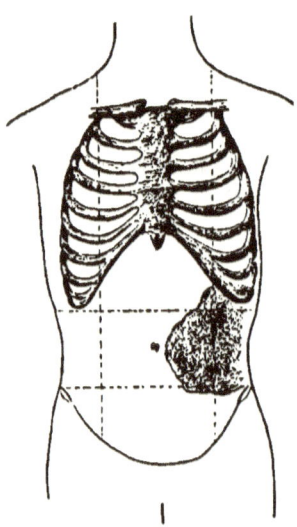

FIG. 41.—Position of the tumor mass in Case LXIII.

Patient menstruated from the 25th to the 28th. The tumor mass was present on the 25th; no examination was made on the 26th. On the 27th the tumor had disappeared entirely. The abdominal walls were so relaxed that palpation could be freely and thoroughly made. The left kidney could be felt on deep palpation. It did not appear to be in any way enlarged; it felt, in fact, rather small and round. A daily note was then made on the patient and the urine carefully measured. The patient says she can always tell a day or so before the attack comes on by feeling dull and the onset of backache.

From October 28th to November 6th the daily note with reference to the left kidney was negative. It was felt every day. She seemed to be doing very well; gained in weight, and had not so much tenderness.

November 7th.—Last night patient had a heavy feeling in the abdomen after eating, and a little distress in the back, as if an attack might be coming on. This morning, however, she felt well again got up, went to town on a street car, walked about a good deal. On her return after dinner the usual symptoms ushering in an attack appeared—slight headache and feeling of sluggishness, and a dull, gnawing ache in the left side, with a feeling of fullness. Patient expresses it that she is entirely "taken possession of by the occurrence," is listless, and if it comes on while she is up and about her knees tremble under her and she feels that she must lie down. She never, however, is nauseated or sick at the stomach.

An examination was made at 11.30 A. M. on the 6th, and the following note dictated : "On drawing a deep breath the left kidney feels a little larger and more prominent than previously, but is not tender." To-day (7th) the examination was made at 3 P. M. The abdomen is slightly distended; the left side more prominent than the right. The tumor mass previously existing again occupies the entire left flank, extending anteriorly almost to the level of the umbilicus. The anterior border is hard, somewhat abruptly defined, and a depression can be felt along the margin. On deep inspiration the mass descends, the lower end almost reaching the anterior superior spine. During the deepest inspiration the fingers can be passed above its upper margin, and the tumor can be held entirely below the level of the tenth rib. From behind, the postero-lateral surface of the tumor is somewhat irregular. With the right hand in the renal region behind it can be readily pushed forward so as to cause a prominent bulging in the left half of the umbilical region. The right kidney is readily palpable and presents no change.

8th.—The tumor mass is not nearly so large this morning. It is firm, rounded, readily palpable between the two hands, and is still large enough to be made to project beneath the skin when lifted from behind. She says she has not nearly so much uneasiness and distress in the side to-day.

9th.—The mass is smaller than yesterday. She has now no pain.

10th.—The tumor has disappeared. The left kidney is readily palpated; feels smaller than the right. A careful estimate was made of the quantity of the urine each day, and the total solids, the reaction, and the specific gravity. From the 28th of October to the 7th of November, during which time there was no tumor, the amount of urine ranged from 1,000 to 1,900 c. c. For the twenty-four hours ending November 7th the amount was 1,900 c. c. On the 8th there was only 1,100 c. c.; on the 9th, 820 c. c.; on the 10th, 1,200 c. c.; on the 11th, 1,210 c. c.; on the 12th, 980 c. c. The urine has always been clear, is usually acid; the specific gravity ranges from 1·010 to 1·017; generally yellow, straw-colored, and contains a few leucocytes. There was no special change in its appearance or microscopical characters, either on the 7th, 8th, or 9th, when the tumor was present, or the 10th, 11th, and 12th, after it had disappeared. The patient went to her home on the 11th. She had subsequently kept account of the amount of urine, which has ranged from two and a half to five pints daily. She had an attack in which the tumor was present on the 17th and 18th, on which days she passed two and a half and three pints of urine, and on the 19th, 20th, and 21st there were only two, three, and three pints. On December 1st and 2d there was again an attack with the tumor present. The amount of urine was three pints on both days, and on the 3d, 4th, and 5th it was three, three, and five pints.

December 16th.—This patient was seen last to-day. She has been better in many ways, but for the past week has not been feeling at all strong, and has been very nervous. The tumor has been present, she states, for about two days. On examination the tumor mass was distinct, though small in comparison with the previous notes. It extended as far forward as the parasternal line, and could be readily moved on bimanual palpation. On deep inspiration the fingers could be pressed above it, and it can be held down. It was distinctly lobulated. In the anterior axillary line it felt superficial, and it could be made to bulge beneath the skin.

A carefully adapted pad and bandage have given much relief, and the attacks have not recurred so frequently. She has also gained in weight and is in every way better.

Additional Note, February 17, 1894.—She has been very much better since last note, and has only had three attacks; one severe, requiring opium. In all three the tumor mass, however, was present. There has been no attack for four weeks, the longest intermission which she has had for months. Her appetite is good, and she now weighs one hundred and twenty-two pounds.

These cases have the following points in common: The patients have borne children; there have been attacks of colic-like pain in the left side, during which a tumor develops, to disappear in the course of a few days, sometimes with an increase in the amount of urine. The diagnosis seems perfectly clear. There is no other condition in which a tumor in the flank appears and disappears in this way. Intermittent hydronephrosis, as is well known, constitutes the most remarkable form of phantom tumor; to-day you may find the side of the abdomen occupied by a large, firm mass which you can grasp between the hands, and which may be so prominent as, when pressed forward, to lift the skin of the abdomen in the region of the navel, and to-morrow you may be completely nonplussed to find that the tumor has disappeared, leaving not a trace behind. There are remarkable cases in which this history repeats itself throughout a series of years, as in the case of congenital hydronephrosis to which I referred—a young man, aged twenty-one years, who had had from his second year the intermittent development of an enormous abdominal tumor which disappeared with the passage of a large quantity of urine.

The subject is one to which much attention has been

given of late, and you will find in the monograph of Landau, and in the works of Morris and Newman, excellent descriptions of intermittent hydronephrosis. The whole question has been most thoroughly considered in the monograph which I here show you,* in which the authors have collected from the literature seventy cases. I see that there has been published recently in London a brochure on the subject by Knight, which has not yet reached me.

A large proportion of all the cases are in women, who are the subject of it at least four times more frequently than men, in about the same proportion as they are more liable to movable kidney. The left side is more frequently affected than the right. Of forty-nine cases in the list of Terrier and Baudouin available for analysis on this point, thirty were on the left side and nineteen on the right.

The general symptoms of intermittent hydronephrosis you have gathered from the report of the cases. In the intervals the patient may feel perfectly well, or may have only the mental worry consequent upon the uncertainty of the nature of the trouble. From this cause Case LXIII lost rapidly in flesh. Case LXI suffers much with the nervous features so often associated with enteroptosis. As a rule, and this is an important point in the diagnosis, the health is good, and the patients are very comfortable, experiencing only, perhaps, a sense of weight or dragging in the side, more rarely local or radiating pains. The examination of the side may be negative; more commonly there is a movable kidney, sometimes feeling quite normal, but it may feel small, as in Case LXIII, or swollen, large and tender. There are instances also in which a sac may be felt, presenting indurated areas, or it may be partly filled. The urine is clear and presents usually no ab-

---

* *De l'hydronéphrose intermittente*, par Félix Terrier et Marcel Baudouin. Paris, 1891.

normal ingredients; in some cases there is a slight turbidity from pyelitis.

You will have noticed in the reports that the attacks recur with variable frequency. Among the circumstances liable to cause them are sudden and violent exercise, the jarring and jolting of riding and driving, any fatigue, mental emotions, and errors in diet. In Case LXIII the patient assured us that she could at any time bring on an attack by a ride in a jolting street car. It is important to bear in mind that indiscretions in eating may cause them. The patient of Dr. Palmer Howard's could at any time bring on a severe renal crisis by taking a heavy supper and a bottle of Bass's ale. The onset is usually manifest to the patient by pain and uneasiness in the affected side and general restlessness. In Case LXIII the patient knew at once when the tumor was developing by the gnawing ache in the left side, the slight headache, and the feeling of sluggishness. The attack may have the severity of nephritic colic and require morphine for its relief. There is rarely fever, nor do I see any cases reported with recurring chills, the absence of which is somewhat remarkable, considering their frequency in affections of the pelvis of the kidney. Nausea, vomiting, diarrhœa, and distention of the abdomen may be present. The attack may last from a few hours to the greater part of a day; the pain gradually passes away, and the patient feels only a soreness and heaviness in the side. The tumor gradually develops during the attack, and may increase in size for several days after the intensity of the pain has subsided. The three patients who have been under observation had learned to recognize the tumor, and knew at once when it was present. In the frequent examinations which I have made of Cases LXII and LXIII I never found them in error on this point.

The tumor itself offers no characters which would call attention to the existence of intermittent hydronephrosis.

It has the situation and relations of a kidney tumor, with perhaps a greater mobility than usually met with in neoplasms or pyonephrosis. When small, it may be very mobile, and some have detected a difference between the renal and the pelvic portions of the sac, separated by a groove. It is deeply placed, rounded, and from behind can be lifted forward from its bed. The median and lower surfaces are smooth, sometimes irregular, but there is no sharp margin or rounded edge. Pressure is often painful, and causes at times an urgent desire to urinate. Fluctuation is rarely obtained, but there is often a sense of elastic resistance. The colon, small bowel, and part of the stomach usually lie in front of the tumor and mask the percussion in the outer half of the umbilical region or in part of the flank.

During the existence of the tumor the amount of urine passed is, as a rule, greatly diminished. After persisting for a variable time the tumor may disappear suddenly with the greatest relief to the patient, and when the evacuation is rapid there is always a notable increase in the quantity of urine. In not one of the three cases which we have considered was the discharge brusque, as in some instances which are on record, but the disappearance of the tumor was gradual, and the increase in the amount of urine, though noted in two of them, was not striking.

With the disappearance of the tumor the patient again becomes quite comfortable, and may remain so for weeks or even months without a recurrence of the attack.

The recognition of the condition, when fully established, is comparatively easy. The pains, the development of a tumor in the flank, its disappearance, usually with an increase in the amount of urine, form a symptom group sufficiently characteristic. It is by no means so easy to determine always the cause. Some of the cases, as already mentioned, are congenital, and have persisted for years. Terrier and Baudouin divide the cases of acquired intermittent hydronephrosis into those in which the cause is obscure or ill-determined, the cases due to lesion of the bladder or of the parts in the vicinity of the lower end of the ureter, and the cases associated with displaced or movable kidney, and due to lesions in the upper extremity of the ureter. In the first group there are a certain number of cases in which the intermittent hydronephrosis is due to calculus. There are instances also caused by blood clots, by tuberculous lesions, and by spasm of the ureter. In the second group of cases the lesion of the bladder is most commonly tumor, with infiltration of the wall near the orifice of the ureter, and by lesions of the uterus and vagina, particularly cancer.

The important rôle in intermittent hydronephrosis is unquestionably movable kidney, the association with which has been recognized since the publication of Landau's monograph in 1881. You will find in the work of Terrier and Baudouin the records of the autopsies which have been made, and of the examinations of the kidneys which have been removed by operation, and I show you here several of the figures which illustrate the marked kinking at the upper part of the ureter. In other instances the ureter has penetrated the pelvis at a very acute angle; and in other cases, again, there appears to have been a positive flexion or twist. It is not difficult to understand how, in the displacement of the organ, such a flexion or kinking could occur, and the wonder, indeed, is that it does not occur more commonly.

You will naturally ask, What becomes of these cases? It is quite possible that the condition may be transient, even when associated with movable kidney. The careful adaptation of a bandage and pad may give great relief, as in Case LXIII; also in case LXII to a less degree. When the attacks are severe and the tumor recurs with frequency, nephrorrhaphy should be urged. The chief dangers are the conversion of an intermittent into a permanent hydronephrosis, and the infection of the sac with pyogenic organisms—conditions which demand operative interference. It is interesting to note, however, the prolonged period during which the contents of the sac remain clear. In the congenital case of twenty years' duration, to which I have so frequently referred, the secretion of the affected kidney—dilated to a shell, but still containing renal tissue—was only a little turbid.

III. MALIGNANT DISEASE.—Of three cases which came before me for diagnosis, two were in children under the age of ten.

CASE LXIV. *Gradual Development of an Enormous Tumor in Left Side of Abdomen.*—E. R., a boy of ten years, seen with Dr. Hewetson, May 10, 1893. I had seen him about six or eight months before for a few minutes in the dispensary. Unfortunately, the notes made at that time have been mislaid. He has had for nearly a year a progressively increasing tumor in the left side of the abdomen. Until two months ago he has been able to get about by himself, but he is now so weak and the tumor is so large that he is scarcely able to walk. He has become very much emaciated, and since I first saw him the tumor has increased greatly in size. His chief complaint at present is of pain down the left leg, and about the ankle and hip of the same side. The abdomen is greatly distended, particularly on the left side, and the superficial veins are very full. The lower part of the thorax on the left side is much expanded, and the costal margin averted.

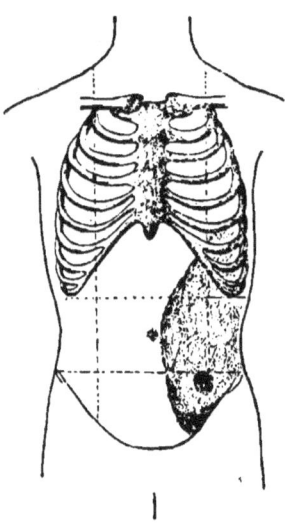

FIG. 42.—Outline of the tumor in Case LXIV.

The entire left half of the abdomen is occupied by a solid mass, which extends from the ribs to the pubes, and is extremely firm and immobile. The surface is smooth, except toward the lower and right margins, where it is irregular. Percussion over it is everywhere flat. It extends upward as far as the seventh rib, and behind to the angle of the scapula. Just above the anterior superior spine of the ilium there is a prominent bulging. The inguinal glands are enlarged and hard, and the supraclavicular glands on the left side are also slightly enlarged. The left leg is slightly swollen, and the veins are enlarged and prominent, as are also the veins over the flank and buttocks. There have never been any urinary symptoms; he has not passed any blood, but he has had pain lately owing to pressure upon the nerves and veins in the left side of the pelvis.

CASE LXV. *Large Tumor in Left Side of Abdomen; Hæmaturia; Removal of Sarcomatous Kidney; Recovery.*—Minnie H., aged five years, admitted to Ward G from the dispensary, January 18, 1893, with tumor in the abdomen, which the mother said had been noticed for several months. She had been failing in health, losing weight, and on several occasions had passed blood in the urine.

Present Condition.—Not greatly emaciated; not particularly anæmic. Veins look blue and the blood a little watery. She has had no fever; pulse quiet. Abdomen greatly distended, nusymmetrical; the greatest prominence is in the hypogastric, umbilical, and left inguinal regions. The whole of the left side is fuller and larger than the right, and does not display the respiratory movements. The superficial epigastric and mammary veins are much distended. There are two prominences to the left of the navel, and one, less prominent, seven centimetres below the costal margin in the left nipple line. There is a fourth just at the tip of the eleventh rib in the left flank. On palpation, the greater portion of the abdomen is occupied by a firm, solid growth, which fills completely the left half, extends fully five centimetres beyond the middle line, and fills the greater portion of the hypogastric region. In the left half of the epigastric region it is covered with stomach or bowel and is not so prominent, and can only be felt on deep pressure. It passes under the costal margin of the

186   THE DIAGNOSIS OF ABDOMINAL TUMORS.

seventh rib. It is extremely firm, resistant, not very movable on bimanual palpation, and not sensitive. It fills the entire flank, and behind is superficial, and gives the impression of occupying closely the whole lumbar region. The surface is irregular, and on palpation the prominences referred to are very distinct. In the right iliac fossa a soft mass can be felt, which is probably the colon pushed over, partly adherent, and can be felt on the tumor mass. Above, as already mentioned, the tumor is covered in the left half of the epigastric region by stomach and intestines, and there is a soft, movable mass, which may represent the curled and thickened omentum.

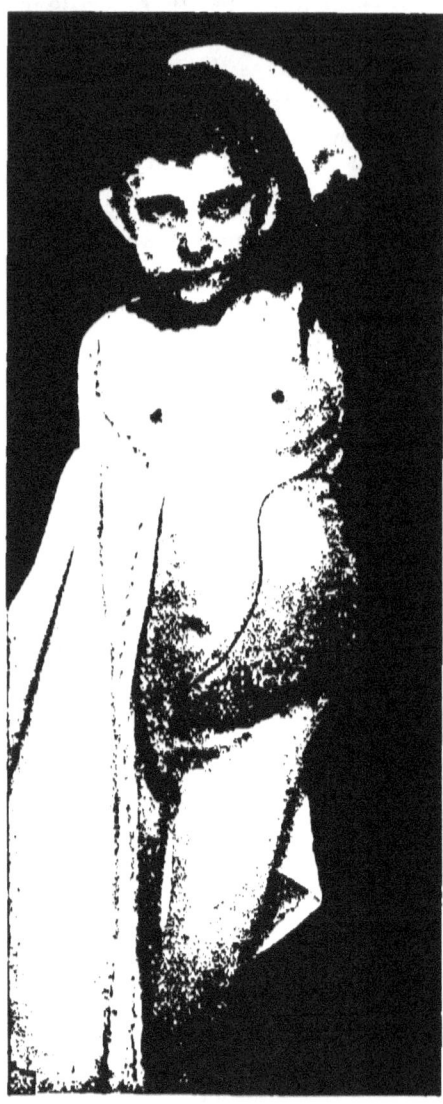

FIG. 43.—Outline of the tumor mass in Case LXIV.

I will read to you the remarks which I made in ward class after demonstrating this child, and which I find here with the type-written report of the case:

"In children, massive tumors of the abdomen are not uncommon, and, as a rule, are either sarcomata of the kidney or of the retro-peritoneal glands. The kidney tumors are the most frequent. Both ultimately produce large, solid growths, which may occupy the greater portion of the abdominal cavity. In the differentiation of these two forms we rarely have any difficulty. Both develop painlessly, and the child may make no complaint whatever; the general health may not be seriously affected, even when the mass has attained a considerable size. Death, indeed, may occur, as in a remarkable case which I have reported of embolism of the heart (the transference of sarcomatous thrombi from the renal vein), before there were any symptoms to attract attention. Progressive emaciation, with enlargement of the abdomen, usually painless, as in the case of this child, are the prominent characters, which are common, however, to both the renal and the retro-peritoneal growth. The two important points of differentiation are, first, the retro-peritoneal growth is more central in its origin, and, if seen early, it is found to occupy the umbilical region, not extending to the flanks; whereas, in the renal tumor, as in the case before us, the growth is lateral, and fills the entire flank, extending deeply behind.

"The kidney tumor is, as a rule, associated with changes in the condition of the urine. Blood is present, either as free hæmaturia, or the constant presence of a small number of red blood-corpuscles. There may be large clots, the passage of which causes great pain. In some cases molds in blood of the pelvis of the kidney and of the ureters are passed, though this is not so common in children as in adults. Other conditions which have to be differentiated are ovarian tumors, pyonephrosis, and cysts, but in cases of doubt the exploratory operation should be strongly urged."

The tumor in this case was removed by Dr. Halsted and found to be an enormous sarcoma of the kidney. The operation was not at all difficult, the child made an uninterrupted recovery, and when last heard of, a month or two ago, remained well.

CASE LXVI. *Large Tumor in the Left Side of the Abdomen; Recurring Hæmaturia.*—November 2, 1893. I saw to-day, with Dr. Lillian Welch, Mrs. X., aged sixty-three years, who had for many months a progressively enlarging tumor of the abdomen with hæmaturia. Until within the present year she has always been a very healthy woman. She has been gradually failing in strength, and within the past six months has lost a great deal in weight, and lately the emaciation has become very great. On several occasions during the past four months she has passed bloody urine, and only at these times has she had any pain.

The patient is a small-framed, much-emaciated woman. The abdomen is distended, particularly on the left side, which is occupied in its whole extent by a large, solid tumor. To the right it extends beyond the middle line, and reaches below the anterior superior spine. It is firm, but with bimanual palpation it can be moved slightly from side to side. The right surface and lower border present large irregularities. It is everywhere flat on percussion, except at the right border. The glands are not enlarged; the superficial veins are only slightly prominent; the examination of the other organs is negative. The urine at the time of my examination was clear. At intervals, however, she has passed considerable quantities of blood, and on these occasions there has been a good deal of pain. Considering the solid nature of the growth, the occurrence of hæmaturia, and the rapid emaciation, the diagnosis of malignant disease of the kidney was thought to be quite clear, and it was not deemed advisable to put her to the pain of an aspiration; nor did her condition seem favorable for an exploratory operation.

The patient died a few weeks after my visit, and Dr. Welch tells me that the post-mortem showed an enormous new growth in the kidney, with small secondary nodules in the liver.

There are two points which you must ever bear in mind in the diagnosis of large tumors in the flank: first, the importance of thorough and systematic examination of the urine with a view of determining the presence of pus, tubercle bacilli, or blood; and, secondly, the use of the aspirator needle. The condition which is really most apt to cause error is the progressively enlarging kidney of pyonephrosis. Gynæcological records indicate how frequently this tumor leads to error, but the chances are reduced to a minimum if attention be paid to the two points I have just mentioned. Catheterization of the ureters may also give information of the greatest value.

IV. TUBERCULOSIS.—A large kidney tumor is rarely due to tuberculosis of the substance of the organ, but tuberculous pyelitis may lead to considerable enlargement of the pelvis and calices, and a certain number of all cases of pyonephrosis have this origin. The tuberculous kidney, however, rarely forms a large abdominal tumor. The following case illustrates two important features in the diagnosis of renal tuberculosis; namely, the determination, by catheterization of the ureters, that the pus came altogether from one side, and the detection of tubercle bacilli in the urinary sediment.

CASE LXVII. *Cough for Five Years; Pulmonary Tuberculosis; Enlargement of the Right Kidney; Pyuria; Tubercle-Bacilli in Urine.*—Susan S., aged sixty-three years, admitted June 4, 1893, complaining of pain in the abdomen. The patient was under observation for a few days in September, 1890, when she had slight cough, and pus and albumin in the urine, but no tubercle bacilli were found at that time.

The patient's mother died of tuberculosis. She has had five children, all living and well. She had pneumonia when thirty-five years of age. For five years at least she has had cough, with slight expectoration, and she has had at times severe chills, which have been supposed to be due to malaria.

Lately she has had a great deal of pain and uneasiness in the abdomen, and has been under treatment for cystitis, though she has had no special pain in passing water.

Present Condition.—Moderate emaciation; marked patchy pigmentation on the face; slight anæmia; no fever; pulse 92, of fair volume. At the left apex the percussion is a little higher in pitch, and there are piping and moist sounds in the first and second spaces. The sputum is muco-purulent, with yellowish lumps, and in these a few tubercle bacilli are found; no elastic tissue.

The heart sounds are clear.

The abdomen is flat, soft, no pain on pressure. The right kidney feels two or three times as large as a normal organ. On deepest inspiration it does not come down far enough for the fingers to be placed above it. No cord-like mass to be felt in the course of the right ureter. The left kidney is not palpable. The urine is turbid, light yellow in color, specific gravity from 1·007 to 1·010, and on settling deposits a creamy pus. There is a trace of albumin, and one or two granular casts are found. Many examinations were made for tubercle bacilli. At first none were found, but subsequently they were found after centrifugalizing the urine, and once in considerable numbers.

On July 12th, under chloroform anæsthesia, Dr. Kelly catheterized the ureters. From the left a perfectly clear urine flowed; from the right, a yellow-brown pus, in which tubercle bacilli were detected. The patient had very slight fever, no chill while in hospital, and appetite and general condition improved very much.

There are two points in the diagnosis of tuberculous pyelonephritis which are well illustrated by this case. The condition is very frequently mistaken for cystitis, and in men more frequently than in women there is great frequency in micturition and great irritability of the bladder, for which on more than one occasion I have known perineal section to be performed. The urine, however, is as a rule acid in tuberculous pyelonephritis, as in this case, unless there is extensive co-existing tuberculous cystitis. The other point is the association of recurring chills. You will have noticed in the history that this patient was supposed to have malarial disease on account of the severe chills which occurred at intervals during the past five years. They may form a very special feature in the disease, as was pointed out many years ago by Owen-Rees, and it is to be remembered that the chills may occur with a very slight amount of pus in the urine.

One of the most important advances in the diagnosis of renal affections has been the facility with which of late surgeons have practiced catheterism of the ureters. Such a demonstration as we had in this case by Dr. Kelly—the catheters in position in both ureters at once, from the right of which a turbid, purulent urine flowed out, from the left a perfectly clear—illustrates the remarkable technique which has been developed by specialists. The demonstrations which many of you have seen in the genito-urinary department by Dr. James Brown prove that catheterism of the male ureters, though not so easy, may be performed with readiness, and gives information of the greatest value as to which kidney is involved.

In the series of cases which we have studied together you have had many illustrations of how far the reasonable probability of Bishop Butler will carry the clinical physician in his endeavors to determine the nature of an abdominal tumor. You will have noticed in how many cases the surgeon made it a certainty, not, unhappily, in diagnosis only, but also in prognosis. But desperate cases require desperate remedies, and in no single instance were the chances of a patient damaged by the exploratory incision.

Amid many pleasant memories of Berlin, just twenty years ago this session, none recur more persistently than those associated with that true Asclepiad, Ludwig Traube, who, adding *probity* to learning, sagacity, and humanity, reached the full stature of the Hippocratean physician. When acknowledging some error he would say—often in a soft, meditative manner, as if gently reproaching himself—

Have we carefully observed all the facts of the case? Yes. Did the art permit of a judgment on the facts under consideration? Yes. Did we reason correctly upon the data before us? No. *Wir haben nicht richtig gedacht.* And with these significant words—may they long echo in your ears!—let us close the exercises of the session.

THE END.

THE
NEW YORK MEDICAL JOURNAL.

A WEEKLY REVIEW OF MEDICINE.

EDITED BY

FRANK P. FOSTER, M. D.

THE PHYSICIAN who would keep abreast with the advances in medical science must read a *live* weekly medical journal, in which scientific facts are presented in a clear manner; one for which the articles are written by men of learning, and by those who are good and accurate observers; a journal that is stripped of every feature irrelevant to medical science, and gives evidence of being carefully and conscientiously edited; one that bears upon every page the stamp of desire to elevate the standard of the profession of medicine. Such a journal fulfills its mission—that of educator—to the highest degree, for not only does it inform its readers of all that is new in theory and practice, but, by means of its correct editing, instructs them in the very important yet much-neglected art of expressing their thoughts and ideas in a clear and correct manner. Too much stress can not be laid upon this feature, so utterly ignored by the "average" medical periodical.

Without making invidious comparisons, it can be truthfully stated that no medical journal in this country occupies the place, in these particulars, that is held by THE NEW YORK MEDICAL JOURNAL. No other journal is edited with the care that is bestowed on this; none contains articles of such high scientific value, coming as they do from the pens of the brightest and most learned medical men of America. A glance at the list of contributors to any volume, or an examination of any issue of the JOURNAL, will attest the truth of these statements. It is a journal for the masses of the profession, for the country as well as for the city practitioner; it covers the entire range of medicine and surgery. A very important feature of the JOURNAL is the number and character of its illustrations, which are unequaled by those of any other journal in the world. They appear in frequent issues, whenever called for by the article which they accompany, and no expense is spared to make them of superior excellence.

Subscription price, $5.00 per annum. Volumes begin in January and July.

PUBLISHED BY

D. APPLETON & CO., 72 Fifth Avenue, New York.

WORKS ON DERMATOLOGY AND DISEASES OF THE GENITO-URINARY ORGANS.

KEYES. A Practical Treatise on the Surgical Diseases of the Genito-Urinary Organs, including Syphilis. Designed as a Manual for Students and Practitioners. With Engravings. By E. L. KEYES, A. M., M. D., Professor of Genito-Urinary Surgery, Syphilology, and Dermatology in Bellevue Hospital Medical College. Being a revision of a Treatise, bearing the same title, by VAN BUREN and KEYES. Second edition, thoroughly revised, and somewhat enlarged. 8vo, 688 pages. Cloth, $5.00; sheep, $6.00.

"Professor Keyes has done the profession good service in this thorough revision of the original work which Professor Van Buren and himself prepared, now many years ago. As the latter states in his preface, litholapaxy has had its birth since that date, the surgery of the kidney has been constructed anew, and very different views are entertained as to the pathology and treatment of many of the abnormal conditions of the genito-urinary system. Thoroughly modernized as Dr. Keyes's important work now is, it will long remain a monument of the skill, originality, and tact of its talented author."—*College and Clinical Record.*

"We can recommend it highly because it is a complete treatise of the diseases of the genito-urinary system, including syphilis, and further, on account of the able and practical manner with which the subject is handled. Any one who will carefully read the pages of this work will find his time has been well spent."—*Canada Lancet.*

VON ZEISSL. Outlines of the Pathology and Treatment of Syphilis and Allied Venereal Diseases. By HERMANN VON ZEISSL, M. D., late Professor at the Imperial-Royal University of Vienna. Second edition, revised by MAXIMILIAN VON ZEISSL, M. D., Privat-Docent for Diseases of the Skin and Syphilis at the Imperial-Royal University of Vienna. Authorized edition. Translated, with Notes, by H. RAPHAEL, M. D., Attending Physician for Diseases of the Genito-Urinary Organs and Syphilis, Bellevue Hospital Out-patient Department, etc. 8vo, 402 pages. Cloth, $4.00; sheep, $5.00.

"We regard the book as an excellent text-book for student or physician, and hope to hear of its adoption as such. In therapeutic detail the recommendations are all good."—*Virginia Medical Monthly.*

"It is scarcely necessary to refer to the talented author of the above-named work, since his life-long labor as a teacher and writer upon venereal diseases has made him known and quoted wherever these affections exist and are treated."—*Polyclinic.*

SHOEMAKER. A Text-Book of Diseases of the Skin. By JOHN V. SHOEMAKER, A. M., M. D., Professor of Dermatology in the Medico-Chirurgical College of Philadelphia. 8vo. With Six Chromo-Lithographs and numerous Engravings. Cloth, $5.00; sheep, $6.00.

"... It is a treatise on the skin which we can recommend to every physician as a work of reference, and in which he will find the latest views on pathology and treatment. At the end of the work are a number of formulæ, which will prove very valuable as a reference. It is certainly a very complete book."—*Canada Lancet.*

"For the general practitioner, this is the text-book he should adopt for practical purposes. Descriptions are good; diagnostic points between diseases bearing a similar phase are clearly made; the drawings well delineate the distinctive features of special diseases; and, above all, the therapeutics suited to given cases are well indicated."—*Virginia Medical Monthly.*

"We shall not be accused of over-patriotic zeal if we recommend Dr. Shoemaker's book as the preferable one for American physicians and students. His style is simplicity itself, and therein has he scored his greatest success over other authors we might name."—*Medical Record.*

New York: D. APPLETON & CO., 72 Fifth Avenue.

www.ingramcontent.com/pod-product-compliance
Lightning Source LLC
Chambersburg PA
CBHW020925230426
43666CB00008B/1572